P9-DFN-544

SCIENCE AND THE REVENGE OF NATURE

MARCUSE & HABERMAS

SCIENCE AND THE REVENGE OF NATURE

MARCUSE & HABERMAS

C. Fred Alford

UNIVERSITY PRESSES OF FLORIDA
University of South Florida Press/ Tampa
University of Florida Press/ Gainesville

UNIVERSITY PRESSES OF FLORIDA is the central agency for scholarly publishing of the State of Florida's university system, producing books selected for publication by the faculty editorial committees of Florida's nine public universities: Florida A & M University (Tallahassee), Florida Atlantic University (Boca Raton), Florida International University (Miami), Florida State University (Tallahassee), University of Central Florida (Orlando), University of Florida (Gainesville), University of North Florida (Jacksonville), University of South Florida (Tampa), University of West Florida (Pensacola).

ORDERS for books published by all member presses of University Presses of Florida should be addressed to University Presses of Florida, 15 NW 15th Street, Gainesville, FL 32603

Printed in the U.S.A. on acid-free paper

Parts of chapter 9 have appeared in *Social Research.*

Library of Congress Cataloging in Publication Data

Alford, C. Fred.
Science and the revenge of nature.

Bibliography: p.
Includes index.
1. Science—Philosophy—History—20th century.
2. Marcuse, Herbert, 1898—Science. 3. Habermas, Jürgen—Science. I. Title.
Q174.8.A44 1985 501 85-627
ISBN 0-8130-0817-4 (alk. paper)

*To Elly, who makes it all more worthwhile, to Thomas C.,
who helped show me the real meaning of reconciliation with nature,
and to my mother and father.*

CONTENTS

PREFACE

THE IDEA of this book began several years ago with an insight I could not then substantiate: that Herbert Marcuse's "new science" was not as unreasonable as it appeared at first glance. I think it was the purely utopian and thoroughly playful character of Marcuse's speculations that attracted me. Since that time, my interests have gone in several different directions, one or two of which are also reflected in this book. After several years of thinking about these issues, I understand better why Marcuse's "new science" has not been received more responsively. I have also come to appreciate the enormous energy and sophistication with which Jürgen Habermas has pursued his project. In the end, though, I would say that my preference for Marcuse's philosophy of science over Habermas' has not changed. What has changed is the enormous number of qualifications, stipulations, and additional considerations that I now believe must be brought to bear on this judgment. In a sense, that is what this book is about. It is also about why the willingness to experiment with utterly new categories of experience, as Marcuse does, is worthwhile.

I could hardly ignore the intellectual background and historical sources of the thought of Marcuse and Habermas, and I certainly do not do so. Indeed, I argue that the problem situation established by Max Horkheimer and Theodor Adorno's *Dialectic of Enlightenment* is never far from either Marcuse's or Habermas' studies of science. However, my study is not an attempt to explain Marcuse's and Habermas' thought by locating their works within the Continental philosophical tradition that runs from Kant to Weber, and beyond (though of course some discussion of this intellectual background is necessary). At the center of my argument is rather an attempt to evaluate the positions of Marcuse and Habermas by pursuing the confrontation between the Anglo-American philosophy of science and Continental philosophy, a confrontation begun by Marcuse and developed by Habermas. This confrontation is also part of the historical and intellectual context of their work, a fact that is appreciated perhaps less frequently than it should be.

William Galston and James Schmidt, two of my teachers at the University of Texas at Austin, rarely let me get away with a sloppy formulation of an issue. I trust this book reflects their influence in this regard. A Fulbright Fellowship to the Federal Republic of Germany allowed me to study with Hans Albert and Helmut Spinner at the Universität Mannheim. I believe I understand the philosophy of social science better for this experience. Jürgen Habermas graciously answered a number of questions I put to him, especially regarding his understanding of science. This was a helpful and enlightening experience for me, though I am sure that Professor Habermas would disagree with a number of my formulations of his position. Trent Schroyer pointed out the complexity of Habermas' position on "theoretically fruitful knowledge," and Gail Landsman directed me to the literature on medical anthropology. Several students in my graduate seminar on the philosophy of social inquiry at the University of Maryland, College Park, never stopped asking questions about many of the issues in this book. To answer them frequently required additional study, the benefits of which I hope are also reflected in this book. Bill Pitt, who prepared the index, is a professional in every sense of the word.

CHAPTER 1

The Issues Involved

SCIENCE IN the work of Herbert Marcuse and Jürgen Habermas is worthy of consideration for several reasons. First, it is an interesting topic in its own right. Second, their different concepts of science seem to act as lenses through which they view nature and humanity's relationship to it. Humanity's relationship to nature has always played a central role in critical theory. Considering Marcuse's and Habermas' views of science helps in understanding how and why the concept of nature is so central to this tradition. Their works also suggest why certain environmental problems are apparently so intractable. Third, this focus upon their views of science illuminates and clarifies their theories of knowledge. The critique of science, and more generally what is called instrumental reason, is absolutely central to Marcuse's and Habermas' understanding of the scope and limits of human knowledge. Thus, this book's focus upon science is both a way of gaining access to several different aspects of the work of Marcuse and Habermas and a way of showing the relevance of recent developments in the philosophy of science to the issue of humanity's relationship to nature.

The philosophy of science has changed a great deal during the last thirty to forty years, i.e., during the course of the development of both Marcuse's and Habermas' work. Both authors have responded to this change, Habermas somewhat more. I take advantage of this ferment in the philosophy of science, bringing to bear fairly recent developments in this field (as exemplified by the work of Thomas Kuhn, Paul Feyerabend, Richard Rorty, W. V. O. Quine, and Mary Hesse, among others) on the positions of Marcuse and Habermas. In this sense my work is not what has come to be called an immanent critique. I believe that Marcuse and Habermas are mistaken in some of their fundamental assumptions about science.

Marcuse and Habermas both write about science from the perspective of the critique of instrumental reason, which was developed by the intellectual leaders of the Frankfurt School of critical theory, Max Horkheimer and Theodor Adorno. This critique argues that science is part of a larger way of thinking, instrumental reason, whose roots can be traced back at least to the *Odyssey* of Homer, but which came to dominate modern thought during the Enlightenment. Characteristic of this way of thinking is its orientation toward nature, and, ultimately, man, as an object to be overcome, dominated, and exploited. Various quotes from that great popularizer of modern science, Francis Bacon, to the effect that science seeks to "hound," "vex," and "torture" nature in order to gain her secrets, are frequently employed to lend support to this interpretation. Part of my argument, drawing upon the contemporary philosophy of science, is simply that modern science philosophically understands itself in such an open, anarchic, and hermeneutic fashion that the equation of science with instrumental reason no longer fits, no matter how thoroughly this equation is modified.

Another part of my argument examines the projects of Marcuse and Habermas from a perspective much more internal to their own goals and purposes. The question asked is why and to what purpose each modifies the critique of instrumental reason in the way he does. I conclude that while there are contradictory themes in Marcuse's work, his version of the critique of instrumental reason serves to rhetorically soften his call for humanity's utter triumph over nature. In contrast to Marcuse's, Habermas' understanding of science has been received very positively within a number of circles outside of critical theory. Not only does Habermas seem more reasonable but his position on science has generally kept pace with developments in the contemporary philosophy of science. Nevertheless, I argue that in some respects Habermas' view of science is less satisfactory than Marcuse's. The way in which Habermas divides and categorizes types of knowledge (Habermas' strategy for dealing with what he regards as the exaggerated claims of the critique of instrumental reason) may contribute to a certain premature freezing of the limits of human knowledge. Certainly there is no more brilliant, innovative, and fascinating social theorist working today than Habermas. Nevertheless, aspects of Habermas' understanding of science—including Habermas' version of a "new science," the reconstruction of cognitive competencies—are in need of correction if they are not to stand as an impediment to further creative thought about man's relationship to nature and his environment.

Though I have attempted to restrict my focus to the issues mentioned, it has not always been possible. Marcuse's and Habermas' concepts of science and nature touch upon so many aspects of their works that it has been impossible to ignore several related issues. For example, chapter 7 is a study of the relationship between Habermas' critique of scientism and his political program. While addressing such additional issues certainly risks a certain softening of focus, the greater risk—one I cannot be sure I have entirely avoided—is to approach Marcuse's and Habermas' views of science and nature from a single perspective in order to make a point. Their views are subtle, complex, and have changed over time. Much of this book is concerned with unpacking their ideas thoroughly, so that the caveats and qualifications made regarding the conclusions I have drawn will be understandable and will be seen as a central part of the argument.

The Issue at Stake

As the introduction suggests, it is widely held that Herbert Marcuse and Jürgen Habermas differ decisively in their analyses of science. Marcuse, it is noted, flirts with "nature romanticism," in which a new science would discover friendly and helpful qualities in nature overlooked by the domineering attitude toward nature characteristic of ordinary science. For Marcuse the basic structure of science is historically relative. A revolutionary change in social relations could bring with it a revolutionary new science as well. Habermas, on the other hand, assumes that the basic structure of science is given by the objective character of human labor. As a mode of instrumental action, science is in the last analysis not much more than a philosophically sophisticated version of labor. As long as nature itself does not change radically—and Habermas does not expect it to—the basic structure of science cannot change either. Science and labor are part of the same struggle to wrest human existence from a nature that is all too sparse.[1]

It is opportune to reexamine this stereotype. Herbert Marcuse's project is complete, and Jürgen Habermas' recent *Theorie des kommunikativen Handelns* casts his previous criticism of scientism in a new light.[2] In fact, this stereotype captures only one stage—and not the latest—in each of their projects. It is opportune to reexamine this stereotype for another reason. Several friendly critics have recently expressed concern that Habermas' view of science and technology is not conducive to the solution of environmental problems. It pits man against nature too severely. From the perspective of Habermas' system, this argument goes, nature cannot help but be regarded as the object of man's "will-to-power."[3] To this criticism, Thomas McCarthy has added that Habermas' separation of instrumental and communicative action is incompatible with the Marxian thesis that nature is the ground of spirit, and that this constitutes a defect in Habermas' position.[4] These considerations have led few of Habermas' critics to embrace Marcuse's alternative, a fact that itself needs explanation. However, these considerations have contributed to a heightened sensitivity among a variety of scholars to the consequences of a perspective that sees nature only in terms of its value to man and the role of science in such an anthropocentric perspective.

Other contemporary critical theorists, such as Karl-Otto Apel and Albrecht Wellmer, deal with science, and no complete survey of the place of science in critical theory could ignore their works.[5] Yet, in a sense Marcuse and Habermas have the first and last words on the subject, respectively. Marcuse is a first-generation critical theorist. His criticism of science is the most radical of any in the last three decades. While Apel's "linguistic apriorism" appears to subordinate science to language even more thoroughly than Habermas' system, it is Habermas who has changed his system several times to account for criticism raised both by the "old philosophy of science" (e.g., Karl Popper), as well as by the "new philosophy of science" (e.g., Thomas Kuhn). There is much to be learned about the limits of critical theory's criticism of science by observing the evolution of Habermas' work. There is also much to be learned from Marcuse's stubborn "Great Refusal" to accept less than the full promise of science: a world relieved from the burden of labor, in which peace and eros prevail.

The most well-known, and certainly the most dramatic, encounter between Marcuse and Habermas over science occurs in Habermas' "Technology and Science as 'Ideology,'" which is a response to Marcuse's "Industrialization and Capitalism in the Work of Max Weber."[6] In "Industrialization and Capitalism," Marcuse suggests that modern science and technology express a historical "project" that is bound up with the rise of modern capitalism. In explicitly using the language of Sartre's phenomenology ("project"), Marcuse expresses the belief that modern science and technology are historically relative enterprises.[7] Change the economic organization of society, mitigate the "surplus repression" that characterizes advanced industrial societies, and a new science and technology might be possible. Such a new science might discover new laws of nature, and a new technology might extract nature's resources in a less violent manner. There is, as will be discussed in chapter 4, some debate as to how radically Marcuse sees the new science and technology. William Leiss, a former student of Marcuse, argues that Marcuse means little more than environmental consciousness and appropriate-scale technology. However, many agree with Habermas that Marcuse is invoking the possibility, under radically changed social and economic circumstances, of an alternative corpus of scientific knowledge and a new technology to go with it.[8] This possibility, says Habermas, "also directs the most secret hopes of" Walter Benjamin, Max Horkheimer, Theodor Adorno, and Ernest Bloch. It is the promise of the "resurrection of fallen nature."[9]

Against the possibility of a new science and technology, Habermas erects an array of arguments that are central to his life's work. He argues, citing Arnold Gehlen, that the achievements of modern technology are a project of the human species as a whole, which cannot be historically surpassed because they are indispensable to human existence as we know it.[10] Modern science too is part of this same species project and is thus invariant, because it necessarily constitutes nature within the framework of possible technical control. There have been refinements in Habermas' position since his publication of "Technology and Science as 'Ideology,'" which will be examined in chapters 6 and 8, especially. Most of these refinements concern attempts by Habermas to distinguish between how the species (including scientists) encounters nature, vis-à-vis how scientists deal with statements about nature that claim to be true.[11] However, in a recent response to his critics, Habermas has again asserted that the only conceivable cognitive orientation toward nature—i.e., one that yields knowledge, not merely aesthetic satisfaction—is one that sees nature in terms of its potential for manipulation and control.[12]

Habermas' claim that modern science and technology are given, and not subject to historical change, is based upon his cognitive interest theory, which argues that the world is constituted (i.e., made knowable to man) under the horizon of two "species interests": the technical cognitive interest in the control of nature and the practical cognitive interest in communication. These interests capture the two fundamental ways in which the world is known to man: as an object of labor and as a place requiring interaction with others. Regarding the technical cognitive interest, Habermas argues that the human species persists over time only because it is capable of apprehending the world as a place to be manipulated and controlled. This interest, says Habermas, is "quasi-transcendental," a term he admits is less than perfect.[13] By this term Habermas means that the technical cognitive interest stems from the biological need of man to realize his natural drives by extracting his existence from nature. Marx called this man's "metabolism" [*Stoffwechsel*] with nature. However, for Habermas this interest is more than just a biological drive. It fixes how nature can be known by man. The technical cognitive interest stems from man's nature, but it goes on to have a transcendental (i.e., knowledge constitutive) function: it limits how man may know the world. That this interest constitutes nature even as it is constituted by nature gives Habermas' claims about the species interest a unique status: half materialistic, half idealistic. This status has been the subject of considerable criticism, a point that will not be overlooked in chapter 6.

Until recently, Habermas seems to have been the undisputed winner in his encounter with Marcuse. Marcuse's vagueness regarding the new science, coupled with the almost mystical connotation of the "resurrection of a fallen nature" have not helped his case. However, as suggested previously, a certain dissatisfaction with Habermas' hard line to the effect that an instrumental orientation toward nature is the only possible cognitive one has become apparent. Henning Ottmann, especially, has made the case that an instrumental orientation toward nature cannot help but see nature as strictly a means to human ends. Such an orientation is not conducive to the resolution of the fundamental causes of environmental problems. Environmental disruption caused by the insensitive employment of modern technologies will likely, under the influence of the instrumental orientation, be met with greater levels of technological intervention. The preferable approach, suggests Ottmann, would involve a certain letting nature be, which is more likely to stem from a perspective that sees nature as a value in its own right.[14] I will discuss this issue in some detail in chapter 9.

The "Revenge of Nature"

It is important to note a preliminary distinction between two forms of the "revenge of nature," a concept that will be referred to frequently. One might call the environmental crisis the revenge of nature—i.e., the consequence of careless intervention in the ecosphere. However, there is another sense of the revenge of nature that was at the center of the first generation of critical theorists' concerns: the revenge of human nature. In this view, which is developed by Max Horkheimer and Theodor Adorno in *Dialectic of Enlightenment*, scientific and technological progress have required the suppression of those aspects of human nature that long to give themselves over to peace and joy: the erotic and the playful. The cost of such suppression is a terrible anger, which can readily be manipulated by a clever demagogue. Habermas, it will be argued, effectively responds to the charge that his system is so anthropocentric that it cannot come to terms with the environmental crisis. A strictly instrumental orientation toward nature need not lead to environmental crisis. Out of their own enlightened self-interest, men can halt acid rain and perhaps even the clubbing of harp seals.

The revenge of human nature is another story. Habermas' solution, the separation of reason into its instrumental and communicative aspects, is not necessarily an adequate one as far as this aspect of the revenge of nature is concerned. In particular, Habermas tends to equate each and every discussion of humanity's reconciliation with nature with an especially extreme interpretation of reconciliation: what Habermas regards as Adorno's and Marcuse's eschatological quest for humanity's universal reconciliation with plants, animals, and minerals. Consequently, Habermas tends to neglect less extreme interpretations of what reconciliation with nature might mean, especially those having to do with humanity's reconciliation with its own instinctual nature. However, while this neglect limits Habermas' analysis, it hardly implies that even these less extreme interpretations of reconciliation could ever be realized. The issue remains open. It cannot be satisfactorily resolved at present, for reasons which will be considered in chapter 9.

However, a related issue is open to resolution. The dissatisfaction with Habermas' dualistic scheme, expressed in terms of a certain skepticism regarding its ability to adequately account for humanity's relationship to nature, can be fruitfully seen from a slightly different perspective. From this perspective the problem is not so much humanity's relationship to nature but that Habermas' dualistic scheme seems to freeze the growth of knowledge. In particular, Habermas' system downplays the creative freedom with which man constructs his philosophy and science. It seeks in some measure to fix in advance the categories of what man makes: the intellectual artifacts of culture. I employ several examples from the field of medical anthropology in chapter 9 to give content to the claim that valid categories of experience exist that find no place in Habermas' system. These examples can be considered as falsifying (though I use the term loosely) Habermas' claim that only an instrumental attitude toward nature is compatible with theoretically fruitful knowledge.

The Concept of Science

While the encounter between Marcuse and Habermas on the status of science takes the form of a debate over whether man's relationship to nature is necessarily instrumental, it would not be fruitful to examine the place of science in their works solely in terms of this debate. Indeed, part of the difficulty of coming to terms with the place of science in their works is that this debate tends to exclude the consideration of other aspects of science in their works, perhaps because this debate is over so dramatic an issue. Habermas has stated recently that developments in the philosophy of science have not occupied his attention in the last decade.[15] Similarly, although Marcuse makes a number of suggestive comments on the topic in his later works, the status of science was certainly not a major theme of his after the publication of *One-Dimensional Man* (1964). Marcuse and Habermas were both responding in large measure to the claims of an aggressive positivism and behaviorism. Recent developments in the philosophy of science, beginning perhaps with the 1962 publication of Thomas Kuhn's *The Structure of Scientific*

Revolutions, have left some of these issues behind even as they were being raised. Indeed, Habermas now argues strenuously against the relativism of some of the exponents of the "new philosophy of science," such as Paul Feyerabend. (Other aspects of the "new philosophy of science," are eagerly embraced by Habermas.) However, while Marcuse's and Habermas' arguments have not always remained current with the recent philosophy of science, their arguments are couched at such a high level of abstraction that they remain relevant. My focus upon their studies of science is not upon their occasional examinations of rather technical issues within the philosophy of science. Nor is my focus strictly upon the debate over the possibility of a new science, though neither of these issues is ignored. My focus is primarily upon what their views of science reveal about their overall projects—and particularly their views on nature and the environment—that is not revealed by other perspectives on their works.

In this context, it might be asked whether it even makes sense to write about something as manifold as science in the work of Marcuse and Habermas. Science is so many things: a set of historical traditions; the literary and popular representation of these traditions (e.g., Francis Bacon's *New Atlantis*); a series of actual institutions (e.g., the National Science Foundation of the United States); a subculture; an ideology; a form of life; an ideal; a model for the "ideal speech situation" (Habermas) and the tolerance of liberal democracy (Popper); a philosophical movement (e.g., "logical positivism"); part of "technique" (Jacques Ellul); part of the capitalist or socialist industrial system; the producer of theoretical knowledge and thus the impetus to the "post-industrial society" (Daniel Bell); populated by a bunch of intellectual parasites (Feyerabend); the epitome of reason; a method; the organon of erotic self-realization (Marcuse); an instrument; a philosophy; a practice; a profession; a set of departments in the university; a discipline. Science is so many things it makes no sense to define it here. Rather, I focus upon a particular *idea* of science held by Marcuse and Habermas, an idea that stems from Horkheimer and Adorno's *Dialectic of Enlightenment*. This idea suggests that the very nature of modern science, its core concepts, indeed its essence—without which it would not be modern science—is one-dimensional, necessarily instrumental, indeed domineering, in its orientation toward nature. Modern science must be these things, this view suggests, because science is an expression of that fragment of reason concerned with human self-assertion. I show that while Marcuse and Habermas are far more subtle than this in their approach to science, this remains the idea of science that guides their works.

The term "instrumental interpretation of science," however, is not unproblematic. In "Three Views Concerning Human Knowledge," Karl Popper states that instrumentalism "can be formulated as the thesis that scientific theories—the theories of the so-called 'pure sciences' are nothing but computation rules (or inference rules); of the same character, fundamentally, as the computation rules of the so-called 'applied' sciences."[16] Under instrumentalism, theories predict the "disposition" of things to do things (e.g., move, break, etc.). While it may sometimes appear that a dispositional statement, such as "the earth moves," is a descriptive statement, the meaning of statements such as this is fully explicable in instrumental terms: "It exhausts itself in the help it renders in deducing certain non-dispositional statements."[17] That is, there is no excess of meaning in the term "the earth moves" that is not accounted for when we use this statement to make predictions about the solar system and the like. Under instrumentalism, says Popper, the conflict between Galileo and the Church turns out to have been unnecessary.[18] Popper goes on to argue that while theories are *also* predictive instruments, instrumentalism cannot account for the fact that scientists stringently test theories that are in fact accurate predictors in order to find out whether these theories are nonetheless false. For the instrumentalist, a false theory that is an accurate predictor is an oxymoron. Yet, scientists make such determinations all the time.[19]

Popper's argument is crystal clear and would resolve the argument readily if this were what all parties in the debate meant by instrumentalism; but it is not. For the critical theorist, instrumentalism does not refer to the epistemological status of theoretical entities but to the nature of the scientific enterprise itself: that it does not produce the type of hermeneutic understanding that is generated when, after listening to the story of another, I finally say, "Now I see what you mean." Rather, scientific knowledge produces not so much understanding as information. Habermas believes that his claim that science is an instrumental enterprise is made at such a fundamental level that Popper's criticism of the instrumental interpretation of scientific theory, while correct, is irrelevant.

> Popper concentrates on the thesis that theories are instruments. Here he can easily counter that rules of technical ap-

> plication must be tried out, whilst scientific information must be tested. . . . The pragmatic interpretation which I wish to give to empirical-analytical sciences, does not include this form of instrumentalism. It is not the theories themselves which are instruments but rather that their information is technically utilizable.[20]

Precisely what Marcuse and Habermas mean when they refer to the instrumental character of the scientific enterprise is discussed in the following chapters. For now it is important to grasp that each is referring to the place of science in human history and in the hierarchy of forms of knowledge. Neither is referring strictly to the status of theoretical entities.

It was the hope of the Enlightenment popularizers of science, such as Francis Bacon, that the increasing rationality of science, evidenced by its successful ability to predict and control natural phenomena, would slowly migrate throughout the rest of society, making its economics, politics, and culture more rational.[21] With Marcuse and Habermas, one sees two different answers as to why this did not occur. Marcuse follows closely the line of thought laid out in *Dialectic of Enlightenment*, according to which the control of outer nature requires the control of inner nature as well: "The subjective spirit which cancels the animation of nature can master a despiritualized nature only by imitating its rigidity and despiritualizing itself in turn."[22] Around the beginning of this century the goal for which the domination of nature was undertaken was being realized in parts of the Western world. Nature was becoming a decent home for man. It did not give up its fruits willingly, but they could be extracted from nature with less and less labor power, less and less pain. However, as Joel Whitebook expresses it, to the extent that the material preconditions for a free society were created by scientific and technological success, the subjective psychological conditions were being distorted.[23] This is the revenge of internal nature. One can trace Marcuse's development of this argument in *Eros and Civilization, Five Lectures, One-Dimensional Man,* and *Counterrevolution and Revolt*, to mention only his most well-known works. Given such an analysis, the pessimism of the early critical theorists, including Marcuse, is understandable. The psychic costs of material progress are extraordinary, but a renunciation of the achieved level of material progress would hardly be conducive to human freedom either. Furthermore, a reconciliation with nature poses its own risks—the risks of a regressive reconciliation, which slips back into prerational forms of thought. The "volkish" ideology of the Third Reich points out the dangers of such an approach. A progressive reconciliation with nature seems to be the only way out. However, exactly what this would look like was and remains unclear.

Habermas avoids this dilemma. In place of the monistic concept of reason held by the first-generation critical theorists, including Marcuse, he posits a dualistic concept. The domination of nature—its utter disenchantment—is accepted as the price of human freedom from constant labor. The realm of human emancipation is confined—once the material conditions of freedom are met—to relations with other men: the realm of communicative interaction. Human autonomy and responsibility do not require the cooperation of outer nature. They require that the self be free from ideological and neurotic self-deception, and this can be achieved, with difficulty, in interaction with others. As Whitebook puts it: "An emancipated society for Habermas consists in the completion, and not the transfiguration, of the modern [Enlightenment] project."[24] The cost of this project, if indeed it is to be regarded as a significant cost, is the acceptance of the disenchantment of the world.

It is at this level of analysis, a level identical to the modern, enlightened view of the world as subject to man's rational intervention and control, that Marcuse and Habermas both tightly link science and technology. Technology, in this view, is little more than corporalized scientific reason. Conversely, science is but abstract technology (though it must be noted that Habermas loosens this linkage in his recent works). However, one might ask, does not a focus strictly upon science in the work of Marcuse and Habermas therefore involve an abstraction that must misrepresent their intent? Would it not be more appropriate to focus instead strictly upon technology, especially insofar as it is not science per se but technology that contributes to (and might help resolve) environmental problems? The answer is no. Indeed, it is hardly unique to critical theory to focus upon the tight relationship of science and technology. The analysis of this relationship has taken several different and well-known historical forms. One view argues that technology influenced science more strongly than science influenced technology. It is well expressed by J. B. Conant's pithy remark that "science owes more to the Steam-engine than the Steam-engine owes to science."[25] Boris

Hessen's analysis of how the basic themes of Newton's *Principia* had their roots in the technological demands of the time, such as artillery, canal-building, navigation, and iron-mining, is exemplary of this approach.[26] It is, however, more common to argue that the relationship works the other way around. T. H. Ashton, in *The Industrial Revolution, 1760–1830*, argues that "the stream of English scientific thought, issuing from the teaching of Francis Bacon, and enlarged by the genius of Boyle and Newton, was one of the main tributaries of the industrial revolution."[27]

On the other hand, Roland Mousnier argues for the total independence of science and technology during this era.[28] Perhaps the most common view today is that science and technology developed relatively independently until about the 1870s, when businesses such as the chemical dye industry began to utilize theoretical scientific knowledge, in this case organic chemistry. Bacon's literary linkage of science and technology was about three centuries too early in this view.[29] Habermas, however, identifies the science of Galileo (ca. 1600) as marking the era in which science was drawn into "a methodological frame of reference that reflects the transcendental view point of possible technical control. Hence the modern sciences produce knowledge which through its *form* . . . is technically exploitable knowledge."[30] For Habermas, Bacon's literary linkage is right on target (his *Great Instauration* was published in 1620 and *New Atlantis* in 1627).

It is neither uncommon nor uniquely characteristic of critical theory to argue that science and technology partake of a common reason. Stephen Toulmin, for example, argues that by rejecting the equation of rationality with logicality, "we can now see scientific rationality and technological rationality as being very closely related: both being concerned with the manner in which new experiences are mobilized to justify changes in our ideas and procedures."[31] Both are concerned with effective adaptation. Science and technology thus may well exhibit common attributes of reason, though the reason they share is not necessarily and obviously instrumental. Science and technology are woven together (though perhaps not so tightly as some suggest) in a complex pattern. Yet, this does not make it an illicit abstraction to focus upon the concept of science in the work of two authors who believe that science and technology are linked especially tightly. In fact, Marcuse and Habermas often focus strictly upon science. Each makes the distinction between science and technology, as well he should. To focus upon science in their works is to focus upon what each regards as the purest and most abstract form of instrumental reason. It is this concept of instrumental reason, in turn, that links their analyses of science with their views on nature and the environment.

Path of the Argument

I do not survey every work of Marcuse and Habermas bearing upon science. Rather, I focus upon their argumentative strategies: what each is trying to establish in his analysis of science. Marcuse certainly seeks man's reconciliation with external nature. However, another value is paramount: man's absolute freedom from external constraint. Only in such a world can eros, not the rationality of instrumental reason, prevail. Science comes to assume a vital role in this project. It becomes the medium of the abolition of labor and thus the vehicle of absolute freedom. Marcuse embraces the material achievements of modern science and technology more thoroughly than any theorist since the great utopian popularizers of science, Sir Thomas More and Francis Bacon. Doing so, while at the same time seeking man's reconciliation with nature, drives his view of science in two quite different directions at once.

Habermas' strategy is quite different. From his cognitive interest theory to his more recent reconstruction of cognitive competencies, he seeks far less from science: that it, coupled with technology, help humanity secure a decent material existence. Conversely, Habermas seeks to prevent the intrusion of science into those areas of life whose rationality is inherently linguistic, not instrumental. As Habermas has recently put it, the "unleashed functionalist reason of system maintenance disregards the claim of reason inherent in communicative socialization."[32] At the level of social policy, this disregard is best overcome, suggests Habermas, by groups of freely communicating citizens, not by mass plebiscite over which elite group is to run things. At the epistemological level, to which he devotes most of his attention, this disregard is best overcome, suggests Habermas, by demonstrating that language cannot be satisfactorily explained by the "objectivating" sciences. Habermas' dualistic solution to the dialectic of Enlightenment requires that language be an autonomous realm, neither fully explicable by science nor reducible to instrumental action. Though the theoretical structure by which this is demonstrated by Habermas changes throughout the course of his work, from cognitive interest theory to reconstructive science, the goal remains the same.

The next three chapters concern the development of Marcuse's argument, beginning with his 1922 dissertation on the German "artist novel"; the subsequent four chapters concern the development of Habermas' argument; and in the concluding chapter I assess the current status of their projects. I argue that Habermas' argument, while extraordinarily fruitful in provoking debate, may have reached an impasse. I suggest that Marcuse's work contains several valuable hints as to how this impasse might be circumvented. This is, of course, not the same thing as saying Marcuse's project is the solution. Rather, it testifies only that I find attractive Marcuse's utter radicalism, coupled with his willingness to entertain some "outrageous hypotheses." However, before beginning with Marcuse's work, it may be valuable to examine the dialectic of Enlightenment in more detail.[33]

The Dialectic of Enlightenment

Dialectic of Enlightenment was written during World War II and published in 1947. The problem situation it establishes for Marcuse and Habermas has already been stated. *Dialectic of Enlightenment* is Horkheimer and Adorno's attempt to come to terms not just with Western science but with the very concept of reason that gave it birth.[34] Horkheimer's famous 1937 essay, "Traditional and Critical Theory," condemns science for accepting the existing order of things.[35] An explanation of the existent that is not also its criticism, suggests Horkheimer, hardly differs from the approval of the existent. This view conditions their definition of positivism. For Horkheimer and Adorno as well as for Marcuse (and to a lesser degree, Habermas), positivism refers to an approach that accommodates itself—even if merely because it seeks only to explain, not to judge—to objective reality. This definition of positivism is best understood in opposition to what might be called "critical negativism," which never rests content with an existing order that leaves even one genuine human need unfulfilled.[36] Critical negativism subjects the existent to constant criticism while at the same time trying to discover progressive aspects of the existent, which might lead to its transformation. It is in this vein that the critical theorists see even such critics of positivism as Karl Popper as themselves being positivists. Like the debate over instrumental reason, the different camps are talking at quite different levels.

Ten years after the publication of "Traditional and Critical Theory," Horkheimer and Adorno express a much more dismal vision. The *Dialectic of Enlightenment* condemns not merely science but the Western intellectual tradition that understands reason as effective adaptation. The dialectic of Enlightenment seeks to explain how fascism could develop within a nation seemingly the embodiment of enlightened liberal ideals. There must be something terribly shallow and vulnerable about these ideals, Horkheimer and Adorno argue, if they could be displaced so readily by the forces of irrationality. Horkheimer and Adorno's explanation focuses upon the key theme of the Enlightenment: its faith in reason. Their criticism of enlightened reason—referred to as the critique of instrumental reason—builds upon George Lukács' consideration that rationalization, in the Weberian sense of demystification and bureaucratization, has done a great deal to liberate man's productive powers but little to liberate man himself. Indeed, rationalization leads to the reification of social relations. Men and women come to see themselves and others entirely in terms of their role in the productive process rather than as individuals. Human relations are transformed into exchange relations.

Horkheimer and Adorno radicalize Lukács' argument. The reification of social relations—as well as humanity's relationship to nature—is traced back to a flaw at the core of the Western idea of reason itself. This flaw, which Horkheimer and Adorno represent in various ways, is that Western reason finds no midpoint between idealism and materialism. Reason and its objects are divided into two realms. Noble ideals, values, discourse over the good life, and so forth are removed to the abstract realm of the intellect and spirit. Like religion, which is an instance of these ideals, these themes are often applauded in the abstract. However, precisely because these values come to be seen as an expression of our "higher" selves, they are disconnected from the everyday material world. The material world, in turn, is given over to a crass materialism that brooks no opposition to the contingently given.[37] The dialectic of Enlightenment is the division of reason into an abstract formal idealism and a crass materialism.

In order to wrest human existence from nature, it has been necessary, according to the dialectic of Enlightenment, to ignore idealistic reason in practice. In effect, reason becomes identical with instrumental reason. Science epitomizes this strategy, in which the laws of nature are learned only by imitating the rigidity of nature itself. This is the real story of Homer's *Odyssey*, according to Horkheimer and Adorno in *Dialectic of Enlightenment*. [38] Odysseus outwits nature and returns home safely. However, in order to do so he must renounce aspects of his own nature. He must have himself tied to his ship's mast, because he knows he has not the strength to resist the Sirens' call—a call that represents the desire to give oneself over utterly to the desires of one's own more primitive nature. Odysseus is rational enough to think ahead and thus outwit his own nature, his own instinctual needs. But, his sailors—like most men—must have their ears stopped with wax, lest they abandon their laborious rowing altogether. The Sirens episode, says David Held, "symbolizes the mode in which crews, servants and labourers produce their oppressor's life together with their own. . . . Their master neither labours nor succumbs to the temptation of immediate gratification. He indulges in the beauty of the song. But the Sirens' voices become 'mere objects of contemplation'—become art."[39]

The *Odyssey* represents the transformation of comprehensive reason into mimesis (imitation) as the price of survival. The physical experiment in science, in which the researcher subjects his every action to the stringent discipline of experimental controls, epitomizes reason as mimesis. Man was weak and ignorant, and nature was powerful and mysterious. Man learned to master nature, but only by imitating her most rigid and routinized aspects. Reason confines itself to a single task: learning to predict and control the contingently given. In doing so it slowly learns to dominate nature, but the price is renunciation. Man must subject himself to a terrible discipline, in which he must deny those aspects of his own nature that are incompatible with the controls of the scientific experiment, as well as the order and regularity imposed by the factory. Indeed, the discipline imposed by the industrial system is but the latest stage in the scientific conquest of nature. Not only must aspects of human nature be renounced, but the full potential of reason must be denied. That dimension of reason concerned with the potential of things to be more than they are (what I have called critical negativism, and what Marcuse in his early work sometimes calls the distance between existence and essence) is split off as idealism, where it remains impotently confined to higher realms. A reason powerful enough to ensure human survival and comfort in a hostile environment is purchased at the price of "Reason" itself. Originating in human weakness, instrumental reason overcomes nature only by renouncing aspects of human nature, as well as the potential of reason itself. Reason becomes powerful only by becoming an instrument.

Horkheimer and Adorno's is not merely an intellectual exercise. Theirs is an explanation of modern history. As reason becomes an instrument of mind, rather than an expression of how the social and natural world is organized (i.e., rather than an objective principle), it comes to be solely an attribute of individuals whose commitment to the Enlightenment ideals of tolerance and respect for the individual is shallowly rooted, and who feel themselves to be powerless. For while science may be powerful in the modern era, the individual is in many respects less powerful than ever before. His or her life is governed by giant bureaucracies, nation-states, the industrial system—a series of powerful but remote abstractions. Though the feudal peasant had perhaps less control over his life, the sources of constraint were clear: the feudal lord, the Church, bad weather. Furthermore, the peasant had the solace of revealed religion and the security of a definite place in the great chain of being.

Modern man is more powerful as a collective entity but often feels himself individually weaker, more vulnerable, more socially uprooted and lost than the feudal peasant. The power of modern science is not his power. Indeed, he may feel even weaker in his own powers when he compares them to the latest machine. The result is an individual susceptible, especially during periods of economic and social crisis, to mass movements that speak to his needs for belonging and meaning. The outcome of the rise of instrumental reason is an individual eager to respond to a demagogue who promises him the power of belonging to a cause greater than himself. This too is the dialectic of Enlightenment.[40]

Horkheimer and Adorno's analysis is truly radical. They suggest, for example, that the categorization and classification of things, such as lion and lamb, is doubly repressive. Categorization originates in repression, for it reflects the hierarchical structure of social classes. Categorization was a repressive social category before it became a repressive mental one.* Moreover, categorization limits the potential of things to become what they are by forcing them into fixed and limited categories. In this sense too categorization is repressive. Only reconciliation with nature, they suggest, would overcome the aggressive categorizing character of modern reason.

What reconciliation with nature means in their works is not entirely clear. It certainly does not mean man's mimetic identification with mere nature. As Adorno put it in one of his later essays:

> The picture of a temporal or extratemporal original state of happy identity between subject and object is romantic, however—a wishful project at times, but today no more than a lie. The undifferentiated state before the subject's formation was the dread of the blind web of nature, of myth; it was in protest against it that the great religions had their truth content.[41]

Indeed, Martin Jay interprets this passage from "Subject and Object" as demonstrating that Adorno held that "for all the costs of leaving behind man's primal unity with nature, his departure was ultimately a progressive one."[42] However, if this is the case, what

* *Primitive Classification*, by Emile Durkheim and Marcel Mauss, originally published in 1903 (Chicago: University of Chicago Press, 1963), the work cited by Horkheimer and Adorno in support of this claim (*Dialectic of Enlightenment*, p. 21), is hardly compelling. As Rodney Needham points out in his introduction to *Primitive Classification*, Durkheim and Mauss believe that the human mind is too weak to construct complex systems of classification. Therefore, these systems must stem from society: "The first logical categories were social categories" (p. xi). Needham notes numerous objections to this claim, among these being that there are many more forms of society than systems of classification (p. xv); that there are profound similarities in modes of categorization among very different societies (pp. xxv, xxvi); and that if the human mind is assumed to be a system of cognitive faculties, not just a passive receptor, then

> it is absurd to say that the categories originate in social organization. . . . [T]he social "model" must itself be perceived to possess the characteristics which make it useful in classifying other things, but this cannot be done without the very categories which Durkheim and Mauss derive from the model. (p. xxvii)

then does humanity's reconciliation with nature mean for Horkheimer and Adorno? The answer is complex and will be addressed in some detail in chapter 9. For now it must suffice to note that reconciliation in their work seems primarily to refer to man's reconciliation with his memory of nature.[43] What is to be remembered? Primarily, it seems, the sacrifices of aspects of human nature—man's playful, spontaneous, erotic nature—that have so long been required in order to master nature. Horkheimer and Adorno's interpretation of the *Odyssey* is an instance of the remembrance of those aspects of human nature suppressed in the struggle for existence that long to give themselves over completely to peace and satisfaction. Reconciled with his own nature, man might come—even in science—to see nature differently, they suggest: not merely as a thing to be mastered, but a place of eros, peace, and beauty as well. Such a relationship to nature would overcome man's sense of isolation and alienation, so cleverly exploited by National Socialism. However, nature is also a place of thoughtless violence and struggle. This is why the simple unreflective identification with nature is barbaric. It is in this vein that Horkheimer writes that "the sole way of assisting nature is to unshackle its seeming opposite, independent thought."[44] In such a world, Horkheimer obliquely suggests, science might move beyond the stage of being an experimental discipline.[45]

Habermas explicitly rejects the "nature romanticism" of Adorno, his teacher. Indeed, it is the purpose of his recent *Theorie des kommunikativen Handelns* to demonstrate that a "consciousness philosophy," such as Horkheimer and Adorno's, cannot adequately deal with the problem of reification in the modern world. Only a communications philosophy that seeks reconciliation strictly between men and women can overcome reification without drifting into virtual mysticism.[46] Marcuse's "defection" is more subtle but also more profound. His new science, which posits man's reconciliation with nature, serves to conceal how urgently he seeks man's ultimate triumph over nature: the abolition of labor.

CHAPTER 2

Freedom and Labor in Marcuse's Early Works

IN THE following chapters, Marcuse's view of science, especially his new science, is explained by considering the function it serves within the larger context of his work. This is really the only alternative to counting quotes, in which selected passages of Marcuse's work are cited to prove that he really means this or that.[1] The issue of the intellectual sources of Marcuse's new science will not be emphasized, although it will not be ignored either. In one respect my approach is almost entirely internal to Marcuse's argument. I simply ask: "What problem is Marcuse attempting to solve with his new science, and what dilemmas does this lead him into?" While I discuss some of Marcuse's early works quite extensively, I do not argue that Marcuse's earliest concerns, particularly with aesthetics, directed his life's work. Rather, it is these early works, along with *Eros and Civilization*, that most clearly demonstrate the problem that the new science attempts to solve.

The German "Artist Novel" and the Priority of Reconciliation

Barry Kātz is certainly correct to see Marcuse's doctoral dissertation, *Der deutsche Künstlerroman* (1922), as well as Marcuse's *Habili-*

tationsschrift, *Hegels Ontologie und die Grundlegung einer Theorie der Geschichtlichkeit* (1932), as key to his later works. This is so even though Kātz sometimes goes too far in reading back into these early works themes that are developed only later by Marcuse.[2]* While neither science nor instrumental reason is discussed in either of these early studies by Marcuse, his *Künstlerroman* particularly is worthy of a brief review because it establishes the centrality of reconciliation in Marcuse's thought. Marcuse's *Künstlerroman* posits an original unity of art and life, which can be seen in pre-Socratic epic poetry, for example, or in the world of the Vikings. This view, says Kātz, is strongly influenced by Lukács' interpretation of Hegel's *Aesthetics*, in which the epic is seen as the art form expressive of integrated cultures. Where this integration is lacking, art turns inward and becomes the novel.[3] In the *Künstlerroman*, the artist is seen by Marcuse as representing a sensibility that is alienated from the surrounding world.[4] Indeed, throughout Marcuse's dissertation the split between various modes of artistic sensibility and the world is emphasized.

> The disintegration and tearing apart of a unitary life form, the opposition of art and life, the separation of the artist from his environment is the presupposition of the artist novel, and its problem of the suffering and the longing of the solitary figure, and his struggle for a new unity. . . . The relationship between knowledge and artistic genius, knowledge and creation, the unworldliness and longing for life of the extraordinarily perceptive artist, is one of the key themes of the artist novel.[5]

It is the nature of the artist, implies Marcuse, to stand outside society. For the purpose of the artist is beauty. However, the artist's

*An intellectual biography, this book succinctly captures the guiding themes of Marcuse's work. However, Kātz's book is not without flaws, though not all the flaws are relevant here (e.g., Kātz fails to adequately document some of his claims—many of which stem from a series of private conversations with Marcuse—such as his assertion that Marcuse was a pseudonymous editor of a Berlin expressionist monthly during the Weimar period). Perhaps Kātz's major theoretical error is his tendency to read back into Marcuse's earliest works themes that are only developed later, particularly Marcuse's employment of the "aesthetic dimension" as a standard of judgment. Nevertheless, Kātz's focus on the importance of Marcuse's early work seems on the whole quite correct. See my review of Kātz's book in *Journal of Politics* 46 (August 1984): 981–83.

world is dominated by philistine concerns. The artist represents an enchanted world of myths, gods, and demons. But, he lives in the "icy polar darkness" of that disenchanted world described by Max Weber. The artist strives for wholeness. He strives to "achieve a life-form which binds the inner disunity into a new wholeness, resolves the opposition between spirit and sensuality, art and life, the artistic existence and the surrounding world."[6] However, even when the artist appears most successful, this integration and unity is frustrated. The tenth chapter of Marcuse's dissertation analyzes the case of Gustav von Aschenbach, the "master craftsman" of Thomas Mann's *Death in Venice* (1911). As Kātz points out, Mann's earlier novellas had depicted the artist as suspended between two worlds, wishing to be assimilated to the comfortable bourgeois life, but remaining foreign to it.[7] Von Aschenbach, however, appears to have successfully entered into the bourgeois world. He is honored and acclaimed. Selections of his work are included in textbooks for schoolchildren. He no longer appears to stand between two worlds. His art has become an honorable profession, which even the bourgeoisie can understand. He lives in Munich, the city of bourgeois artistic genius, "*in bürgerlichem Ehrenstande*."[8]

However, in spite of its attractiveness to the artist, says Marcuse, this bourgeois way of life can never truly be the artist's own [*ist ihm nicht 'angeboren'*]. Von Aschenbach "belonged to a different order of humanity, to a different world, and in the face of the Dionysian forces which have their roots in that humanity and that world, no heroism or determination could protect him . . . if they break through only once, they demolish the bourgeois existence, shatter all harmony, bring to ruin all stability and order."[9] The world remains split. The artist can never truly be integrated into it until the world too is organized on the principle of these same erotic drives. The artist is condemned to serve the forces of critical negativism.

There is, however, some ambiguity in Marcuse's treatment of Mann's novel. Marcuse suggests that *Death in Venice* represents an exorcism or catharsis [*Beschwörung*] for Mann. The very process of writing the novel allowed Mann to liberate himself from the demonic powers and artistic alienation about which he wrote. This would seem to imply that the artist need not be permanently alienated from the bourgeois world. The artistic process itself, by briefly allowing the artist to enter another world (albeit solely in his imagination), is so cathartic that it facilitates his integration with this world. Yet, this does not seem to be Marcuse's final position. As Douglas Kellner points out, immediately after suggesting that the cathartic experience of artistic creation reconnects the artist to this world, Marcuse goes on to argue that a genuine and permanent connection requires "a 'community' [*Gemeinschaft*] in the most extreme and deepest sense. It alone is the solid and fruitful ground out of which the great epics arose, and on which the nay-saying [*entsagend*] artist can realize an appropriate and satisfying accommodation [*Einordnung*]."[10] It seems that the artistic catharsis is only temporary. It can become permanent only when the world itself reflects the coherence of the artist's vision.

This ambiguity in Marcuse's position does not disappear. In his last book, *The Aesthetic Dimension*, written over fifty years later, Marcuse suggests again that it is through aesthetic catharsis that man comes to terms with the ultimate duality of man and nature, subject and object. Man may bend nature to his purposes through science and technology, but he may not, apparently, make nature his partner. That "the world was not made for the sake of the human being and it has not become more human" is the final regrettable fact that man comes to terms with in the aesthetic catharsis that Marcuse writes of in *The Aesthetic Dimension*.[11] This would seem to imply that Marcuse abandoned the most ambitious aspect of his new science, in which it would recast nature as a helpmate. Yet, even in *The Aesthetic Dimension* an element of ambiguity remains on this score. What I try to demonstrate in this and the next two chapters is how the strategy for freedom that Marcuse adopts leads him to emphasize man's triumph over nature over man's reconciliation with nature, all the while recognizing that both strategies coexist in Marcuse's work until the very end.

Marcuse's study of Mann concludes his dissertation. His analysis of the situation of the artist goes no further. Though Marcuse mentions such things as "the powerful boom of commerce, of industry and technology, [and] the conquest of natural science," these are considered only as background against which the artist reacts.[12] It is upon the artist's alienation, not its environmental causes, that Marcuse focuses. In this respect Marcuse's dissertation differs from his later works. Nevertheless, the themes of his dissertation influence Marcuse's life's work. The most important themes are: (1) the identification of a rent in an original and unitary way of life [*die Zersetzung und Zerreissung einheitlicher Lebensformen, der Gegensatz von Kunst und Leben*]; (2) the heightened sensibility of the artist who detects it; (3) his inability to integrate himself into the bourgeois way of life; and (4) the power of the aesthetic and erotic forces that will prevent permanent integration until this world is once again organized on the basis of these forces. However, it would be a mistake to read Marcuse's later work solely, or even primarily, in terms of how each of these themes is developed, although such an approach has its place. Later influences upon Marcuse, especially those of Heidegger, Marx, and Adorno, do not merely build upon these themes but serve as the impetus for new ones. Perhaps the most enduring aspect of these early themes is found in Marcuse's constant awareness of the potential and danger of a false integration with a merely comfortable world. It is this theme that characterizes so much of his later work.*

Hegels Ontologie, published in 1932, concerns similar themes. It is sufficiently complex and not directly concerned with the topic at hand, so that it can only be briefly mentioned here. Kātz notes that like the *Künstlerroman*, Marcuse's *Habilitationsschrift* is concerned with the problem of overcoming a divided world.[13] In *Hegels Ontologie*, the division Marcuse is concerned with is primarily epistemological: the dualism of subject and object that was the legacy of neo-Kantianism. Marcuse contends that this division, which is actually the characteristic of a particular historical era, was transformed by neo-Kantianism into a given condition of human existence: an ontological fact.[14] Marcuse would, presumably, say much the same thing against Habermas' cognitive interest theory: that it treats historical distinctions, such as that between instrumental and communicative rationality, as transcendental facts given for all time. However, the few published exchanges between Marcuse and Habermas have rarely (with the exception of Habermas' "Technology and Science as 'Ideology'") addressed the fundamental assumptions of either of their systems. Their exchanges have been friendly and in some ways superficial.[15]

In *Hegels Ontologie*, Marcuse argues that Hegel confronts this sit-

*Douglas Kellner, in *Marcuse and the Crisis of Marxism*, argues strenuously against Kātz's reading of Marcuse's *Der deutsche Künstlerroman* as attempting to carve out a realm of ontological stability that is resistant to the contingency of everyday life (p. 384). Rather, says Kellner, Marcuse treats art in his dissertation as being at its best when it expresses "joy in reality, which is integrated with life and is 'affirmative'" (p. 384). Nothing crucial hangs on this disagreement as far as the issue that I am addressing is concerned. The themes that I have characterized as influencing Marcuse's work (Kellner agrees that there are "uncanny anticipations of the later Marcuse in his dissertation," p. 384) do not depend upon whether the unity that Marcuse

uation of division [*Situation der Entzweiung*]—so similar, Kātz points out, to the division of the aesthetic and bourgeois worlds in the *Künstlerroman*—as a problem, to be overcome by philosophy and human action in the development of world history. The "end" of this historical process, expressed, notes Kātz, in the central concept of *Sichselbstgleichheit-im-Andersein* (selfsameness-in-otherness), is the overcoming of the division between subjectivity and objectivity, actuality and potentiality, abstract essence and concrete existence.[16] The pursuit of this realm of freedom in its various forms, from Hegel's earliest theological writings and throughout the *Phenomenology of Mind* and the *Logic*, is, states Kātz, the structural theme of *Hegels Ontologie*.[17]

Whether or not *Hegels Ontologie* is compatible with Marcuse's later treatment of Hegel in *Reason and Revolution* is not at issue here.[18] Neither shall the issue of whether Marcuse adequately exploits the antimetaphysical possibilities of Hegel's thought be considered. Though Adorno, in a review in *Zeitschrift für Sozialforschung*, saw a promising movement in Marcuse's *Hegels Ontologie* away from a concern with the "meaning of Being" and fundamental ontology and toward the disclosure of actual beings and the philosophy of history, he was perhaps being generous. As Kellner points out in *Marcuse and the Crisis of Marxism*, Adorno's "is a strange reading of Marcuse's first Hegel book, since he [Marcuse] rarely mentions concrete history but stays at the level of pure ontology."[19] In any case, what surely is not debatable about this early work is Marcuse's persistent concern with the reconciliation of subjectivity and objectivity, man and world. Once again, as with the *Künstlerroman*, one sees no program as to how this is to be achieved. However, like the *Künstlerroman*'s treatment of the epic, one sees that for Marcuse the potential unification of opposites "presupposes for its part a prior signification which is its ground, an originary synthesis (!) which is the 'measure' of all comparison and all opposition."[20] The unity that it sought in the future reflects an earlier lost unity, which serves as the measure [*Massstabe*] of progress toward reconciliation. This idea of a lost unity that might possibly be recaptured reappears in Marcuse's later works. It plays a role in his new science, a project that is not given explicit form by Marcuse until three decades later, with the publication of *One-Dimensional Man* in 1964.

---

seeks belongs to a higher realm of being. In fact, Marcuse seems to vary his position on this issue, as subsequent discussion will suggest.

Three Early Essays

Not Hegel but Heidegger is usually seen as the dominant influence on Marcuse's early essays. In fact, Heidegger's influence on Marcuse's concept of science is subtle, roundabout, and interesting. Marcuse's first published essay, "Contributions to a Phenomenology of Historical Materialism" (1928), is the subject of considerable controversy.[21] In it Marcuse gives a Marxian answer to the Heideggerian question "What is the essence of human existence?" Marcuse's provocative answer is that human existence is to be understood by the way in which it "cares for itself" in the world, and the primary care of existence is its production and reproduction in labor.[22] Alfred Schmidt calls this an extraordinarily difficult way to reconcile Marx and Heidegger. It is so difficult because recognizing the centrality of the actual material situation of man, such as his need to labor, explodes the Heideggerian emphasis on the subjective experience of history [*Geschichtlichkeit*].[23] A consequence of this view, says Schmidt, is that Marcuse bases the propensity to revolt upon the need of human existence to realize its own essence, rather than upon a concrete analysis of structural factors such as poverty and unemployment. Marcuse states: "Organic historical development and revolution are so uncontradictory that revolution appears rather as a *necessary* form of historical movement; for revolution alone can transform the existence of historical being."[24] Schmidt concludes that historical objectivity becomes merely the material of duty, a "decoration" for a decision no longer truly in the hands of deciding individuals.[25]

Morton Schoolman sees this first essay in quite a different light. In the years before Marcuse had access to Marx's *Economic and Philosophic Manuscripts of 1844*, says Schoolman, Marcuse employed Heidegger's analysis to fill in what seemed to be a gap in Marxism: how a concretely existing individual could develop a revolutionary consciousness.[26] Heidegger's analysis of man's almost natural alienation from a world he finds himself "thrown into" provides the basis for Marcuse's analysis of an individual predisposed to radical action. Should the structural concomitants of revolution be present, the revolution can build upon this existential disposition to radically remake this world in which man finds himself, a world in which he is never completely at home.[27]

Though the extent of Heidegger's influence on Marcuse's early work is debatable, Marcuse's view of science in this first essay is quite clear.[28] This is the primary issue with which I am concerned here. Marcuse sees science in phenomenological terms that appear to be drawn from Heidegger. The scientific perspective, says Marcuse, is based upon a theoretical abstraction that regards the world as "purely present" [*Vorhandenheit*]. However, says Marcuse, this scientific abstraction is not illicit or misleading. Natural science can and must abstract because its object is not historical.[29] The problem of illicit abstraction arises only when those sciences concerned with the essence of human existence and its meaning also abstract from history. Marcuse here asserts a classical division of the natural and social sciences based upon the character of their objects. Surely, says Marcuse, we can have a dialectical analysis of nature insofar as this analysis is concerned with the historical relationship of man and nature. However, this is not the same thing as natural science, and one must certainly not confuse them as Engels did in the manner of a new physics. It is, adds Marcuse, useless to ask which approach to nature is correct, for they both are. They are simply concerned with different classes of objects. Science can treat nature independently of its human significance precisely because the essence of nature is not historical.[30] Nature remains exterior to man and his purposes.

Marcuse posits in this essay—Schmidt points out that this is an epistemological implication of Heidegger's *Sein und Zeit*—a realistic interpretation of the objectivity of the natural world, in which our immediate everyday experience of the facticity of nature is the basis of our knowledge of it, even at high levels of scientific abstraction.[31] Marx and Heidegger may not fit together very well; however, a realistic interpretation of the objects, operations, and theories of natural science is for Marcuse not incompatible with an abstract phenomenology of human historical essence.

In "On the Problem of the Dialectic" (1930, 1931), Marcuse reasserts that the border between the historical and nonhistorical is parallel to the border between human and natural history.[32] Only when the object to be dealt with is itself dialectical can its analysis also be. Dialectics refers to a real process in history, not merely a cognitive act. As in "Contributions," this is understood to exclude a new science or nature. Only human history is genuinely dialectical. However, this position is modified in "Das Problem der geschichtlichen Wirklichkeit: Wilhelm Dilthey" (1931).[33] In this essay Marcuse suggests that there is an even more fundamental stratum of experience in which this distinction, while not disappearing, is itself drawn into history. Marcuse explicates this fundamental level of experience by means of Dilthey's concept of *das Leben* (life).[34] Marcuse argues that in the course of Dilthey's work the concept of *das Leben* changed. At first, *das Leben* was set against the domain of the natural sciences and was defined by the interest of the humanities in meaning and significance.[35] However, as Dilthey became interested in the development of historical—not just individual—life, the concept of *das Leben* began to encompass more and more of experience until it came to include man's experience of nature, rather than being set against it. *Das Leben* is the culture of a society in the broadest sense, and it comes to be such a comprehensive concept that it need no longer be defined by contrast to the concerns of science. It includes science and man's experience of nature. *Das Leben* is the historical context that gives meaning and significance to every human experience. Marcuse writes, "Dilthey came upon the revelation of the historicity of the self-determination of 'das Leben,'"[36] and he goes on to argue that under the identification of the concept of *das Leben* and historicity, traditional (i.e., ontological) distinctions between man and world, nature and history, subjectivity and objectivity, take on a wholly different significance. These distinctions obtain not only their meaning, but also their very existence and form, from the achieved level of historical development —from the form of *das Leben* itself.[37]

There is no hint of the new science in these early essays and certainly not in the *Künstlerroman* or *Hegels Ontologie*. Indeed, in these essays Marcuse upholds a realistic interpretation of science and the natural world it is about, an interpretation that seems to owe almost as much to Heidegger as Marx. Yet, while there is no hint of the new science per se, the foundations of Marcuse's contradictory attitude toward nature are apparent. Nature is held rigidly apart from human history even as Marcuse is developing categories such as *das Leben*, which must subvert this distinction. To be sure, nature is indeed a social category, but it is not only a social category, a point Marcuse's interpretation of the concept of *das Leben* will allow him to forget.

Heidegger also proposed what might be called a new science, characterized by a receptive, waiting-for-being-to-appear philosophical method. Such an approach to nature, Heidegger suggests, might reveal another side to nature, perhaps a friendly helpful side. "It could yet be the case that nature has in fact concealed her essence by turning to men that side which can be technically mastered."[38] Marcuse's new science, it will be shown, exhibits such an attitude toward nature, in which "sensuous perception" would let itself be guided by the pleasant qualities in nature that it discovers. The relationship of this view to the "nature romanticism" of Adorno, particularly, is apparent. A new side of nature would be revealed by remembering the reality constitutive character of *das Leben*.

Nevertheless, Schmidt's interpretation of Marcuse as holding a realistic view of science and nature during these early years is on the whole quite correct. It is not Heidegger's philosophical method, his receptive orientation toward the being of nature, that is the decisive influence on Marcuse's concept of science, but it is rather Heidegger's analysis of man's profound alienation from a world he finds himself "thrown into," an alienation so fundamental it can hardly be overcome by socialist revolution. For this alienation has to do with the physical conditions of human existence: that we are born into this world alone, and we will die that way all too soon, having spent most of our lives in labor.[39] "History is the negation of nature," writes Marcuse. "What is only natural is overcome and recreated by the power of Reason."[40] This is hardly a passive, waiting-for-being-to-disclose-itself attitude toward nature. Rather, Marcuse seeks to overcome nature as thoroughly as possible. In the course of his work this comes to mean relieving man of the historical burden of labor so that he may devote all of his regrettably brief existence to the pursuit of pleasure. Marcuse recognizes that science and technology are the only tools man has available that might possibly allow him to realize this goal of effortlessly overcoming nature's scarcity. At the same time, Marcuse believes that science and technology express a domineering attitude, one that sees in nature only a means toward human ends. Much of the new science seems designed to rhetorically soften this harsh reality: that only by treating nature as a pure means can human freedom be wrested from it.

Freedom and Labor

Marcuse's 1933 essay on labor is the most decisive of all Marcuse's early works as far as the development of his concept of science is concerned. In this essay, science per se is hardly mentioned. However, the discrepancy between man's existence and his essence (i.e., his potential) is stated more concretely than before. Indeed, it is stated so concretely—in terms of man's need to labor—that the discrepancy can be addressed not merely by art and philosophy but by science and technology. The influence of Marx in this regard is apparent.

Why is labor inherently unfree and burdensome? In "On the Philosophical Foundation of the Concept of Labor in Economics" (1933), Marcuse answers that in labor the thing itself, that which is labored upon, always comes first.[41] In labor man must subordinate his will to the requirements of the objective situation and thus cannot labor for himself in the way he can play for himself. Play is the opposite of labor and the epitome of freedom. Play cancels the objectivity of objects and replaces it with humanly created laws: the rules of play. In play natural (scientific) laws are ignored, if only for a moment, and in this period of time man is free to be himself in a way he never can during labor. "In a player's single throw of a ball lies an infinitely greater triumph of human freedom over objectification than in the most enormous accomplishment of technical labor."[42] In play man is free to be himself because he plays only for himself, not for external purposes. His rules, and not those of the object, prevail.

Labor is burdensome not only because man must subject himself to the "law of the thing" [*Gesetz der 'Sache'*][43] but because he must do so constantly and presumably permanently. The objective permanence of labor is based upon natural insufficiency, in which the material of human existence must constantly be extracted from the earth to which it ultimately returns. For this reason the process is permanent, or as permanent as life itself. Marx called this the "metabolism" of man and nature.[44] However, there is another aspect of the permanence of labor, which Marcuse calls the subjective aspect. The subjective permanence of labor is not well defined in this essay, but it seems to refer to the idea that man realizes himself (i.e., his essence), his own potential to become more than he is, by passing through a labor-like cycle of objectification-reification-appropriation. How these themes interact is explained below.

In labor man objectifies reality. The world is seen by man as the field of human action, entirely external to himself, which must be transformed in order to meet his needs. Objectification—the positing of the external world—implies that man must appropriate aspects of this world to meet his needs. In appropriation: (1) the given is transcended, in the sense that man realizes that the given can be changed by being worked upon. (2) In working upon things, man gives them a form and reality that is his own. For example, he turns a tree into a table. (3) Man realizes himself in the thing he has made, in the sense that he sees his own powers to transform things, as well as his own vision of what the things should become, in a congealed form, as it were, in the things he has made. The problem is that objectification tends toward reification. It is as though man's powers to transform things pass over into them without his ever having been aware they were his own powers. Man's powers are thus estranged from himself at the very moment they should have become most real to him. For Marx, this feature of labor can be traced back to the structure of capitalism and thus is not in principle an endless process. It would end under communism. Marcuse, however, states that this process is based on a fundamental lack in human existence itself, in which man can only realize himself by passing through externalization and alienation again and again. The process is endless.[45]

It is not entirely clear why the subjective as well as the objective aspect of labor is presumed by Marcuse to be an endless cycle. It is not clear what the lack in human existence is that requires passing through externalization and alienation again and again. For Marcuse seems to be saying in his discussion of play that man realizes himself most fully by not passing through this endless cycle but by breaking it. Self-fulfillment, and with it the reparation of the lack in human existence, comes not through the processes of objectification-reification-appropriation but in being a world unto one's self. In play man denies the objectivity of the world and puts his own objectivity in its place. In this way, he does not have to appropriate something "out there" to begin with. Objectivity is, as it were, pre-appropriated. In fact, there is an easy way to reconcile this apparent contradiction. Indeed, Marcuse hints at this way, though it tends to get suppressed because Marcuse uses the terminology of human self-realization—such as a "lack in human existence," the negativity of human existence," or "the goal of labor is human existence [*Dasein*] itself"—when referring to both the objective and subjective aspects of labor.[46]

If play is the genuine fulfillment of human existence, it is no wonder that man never fulfills himself in labor. It is no wonder that the subjective aspect of labor would also be permanent. Labor is simply not the realm of self-fulfillment, no matter how it is socially organized. The essential lack in man, which Marcuse quite misleadingly refers to as the subjective aspect of labor, can be traced back directly to nature. Nature is so sparse it does not allow man enough time for free self-development; it does not allow man enough time to play. Insufficient time for self-development is the essential lack in man, a lack directly attributable to nature's sparseness. Man passes through the eternal cycle of objectification-reification-appropriation because it is necessary to maintain his existence, not because it is a Sisyphean task of self-realization. Labor by its very nature can never contribute to the realization of man's existence. In fact, Marcuse closely identifies the duration, permanence, and burdensomeness of labor. Man's subjection to the "law of the thing" (worldly objectivity) is so burdensome precisely because labor is man's primary task. Play is only a brief respite from labor by its very definition. No duration and permanence belong to play in the totality of "human existence: it takes place 'from time to time,' 'between' the times of other activity, which continually and permanently dominate human existence."[47]

In "The End of Utopia" (1967), Marcuse writes of the possibility of "letting the realm of freedom appear within the realm of necessity," and in "Protosocialism and Late Capitalism," written shortly before his death, he expresses this ideal once again.[48]* Yet, this is but a profession of faith and hope. Nowhere, including these es-

*"The emergence of the realm of freedom *within* the realm of necessity" is how Marcuse expresses this ideal in "Protosocialism and Late Capitalism," p. 30. I see no evidence that Marcuse asserts this possibility in any of his pre-1967 works. When, in *Reason and Revolution,* Marcuse refers to the abolition of labor he refers specifically to Marx's concept of labor as "that activity which creates surplus value in commodity production, or which 'produces capital'" (p. 239). Marcuse is referring to the abolition of the capitalist concept of labor and not to man's "metabolism" with nature, which is the concept of labor under consideration here. In "The Realm of Freedom and the Realm of Necessity: A Reconsideration" (1969), Marcuse briefly contrasts the "classical Marxian conception" with that of the young Marx. (In spite of its title, this short essay is primarily about the student movement.) While the later Marx saw no place for freedom in necessity, the Marx of the *Grundrisse der Kritik der politischen Ökonomie* "envisages conditions of full automation, where the immediate producer . . . can play with, experiment with . . . the possibilities of the machine" (p. 22). Though Marcuse does not pursue this point in this essay, one can read *Eros and Civilization* (1954) as an attempt to develop this notion, as will be shown in the next chapter.

says, does he pursue this possibility systematically. The relationship of freedom and labor remains the one established in this essay. They are incompatible. Some have seen this as a fundamental defect in Marcuse's position. For example, Douglas Kellner argues that "Marcuse's early ontology of labour reproduces recurrent features of labour under capitalism which are projected into a universal Concept, thus falling prey to what I call 'the fallacy of ontological generalization.'"[49] The features Marcuse ascribes to labor—externalization, burdensomeness, constant and enforced activity—describe features of wage labor under capitalism, and are not necessarily universal features of all labor at all times. Kellner makes a good point, especially insofar as the alternative is seen as labor in precapitalist agrarian societies, in which "labour often partook of features of play, communal ritual and festival, or was often only seasonal and intermittent, and thus was not a constant task or function, as Marcuse's analysis implies."[50] Yet, it is difficult to imagine how the burden imposed on human freedom by the concept of labor that Marcuse is considering could ever disappear in even the most thoroughgoing socialist society. It is not merely that Marcuse "ontologizes" the concept of labor; rather, he transforms the concept itself.

Marcuse does not simply take features of labor that are characteristic of the activity under capitalism and call them universal. Rather, he extends the concept until it comes to include humanity's interaction with worldly objectivity itself. If throwing a ball is a greater expression of human freedom than the most enormous accomplishment of technical labor, no matter how this labor is socially organized (this last clause is surely the implicit context of the entire article), then we are faced with a way of looking at the world that does more than universalize Marxist historical categories. Though Marcuse's concept of subjective labor is not entirely transparent, it seems to refer to the constraints placed on free self-development by the objectivity of the natural world: not merely the time spent in labor but the limits of the human life-span itself interfere with freedom. It appears that the "law of the thing" has little to do with the social organization of labor and everything to do with the constraints on human self-realization imposed by nature's laws (this point is explored more fully in the next chapter). From this perspective, it is apparent that Marcuse seeks not so much to "ontologize" labor as to "deontologize" its necessity. That is, he seeks to render necessity—including the necessity of nature itself—merely historical. While this program certainly involves Marcuse in a certain "spiritualist contempt for the finite and terrestrial world," as it has been called, what is so striking about Marcuse's project is the degree to which this struggle against worldly objectivity becomes an utterly mundane project.[51] It becomes mundane by virtue of Marcuse's fundamental strategy for freedom, which is to combine the insights of his *Eros and Civilization* with the possibility of total automation made possible by scientific and technological progress. How Marcuse combines these elements is a major theme of the next two chapters.

The preceding considerations suggest the way in which science and technology come to serve an absolute goal in Marcuse's project. In Marcuse's interpretation of Hegel, cognition and action are unfree because they are oriented toward a world that is not yet mastered. Only pure thought is free, for thinking itself has no object but itself.[52] In Marcuse's version science replaces pure thought but the idea remains similar: freedom can be achieved only when all constraints on human action are eliminated:

> Science itself has rendered it possible to make final causes the proper domain of science. . . . The transformation of values into needs, of final causes into technical possibilities is a new stage in the conquest of oppressive, unmastered forces in society as well as nature. . . . *Technology thus may provide the historical correction of the premature identification of Reason and Freedom.*[53] (my emphasis)

One should always be careful with such parallels. They only reach so far, and this one (between Marcuse and Hegel) certainly does not exhaust Marcuse's understanding of the role of science, particularly. Nevertheless, it is a useful parallel because it suggests that Marcuse's concept of a truly emancipatory science is not merely an expression of a passive, receptive, "waiting" attitude toward nature. Far from it. Rather, science and technology do the real work—they conform to the "law of the thing"—so man can realize his freedom beyond this domain.

One can see in the preceding quote how science and technology in the service of an absolute goal can themselves get drawn into this goal, in this case as a science of final causes. How is this to be explained in view of Marcuse's adherence to what in many respects is a traditional, albeit particularly instrumental, view of science? This question will be answered in several ways. However, one of the most fundamental answers is given by Alfred Schmidt, though he is not referring to Marcuse's work, but to characteristics of utopian thought in general.

> The peculiar idea that a fundamental change in the whole universe will go hand in hand with the proper organization of human relations can already be found in the early socialist writers of the pre-1848 era. Fourier's fantasies are moving, for as Benjamin recalls, in Fourier rationally organized labour is supposed to have the result "that four moons light up in the terrestrial night, that the ice retreats from the Poles, that sea-water no longer tastes of salt, and that predatory animals become the servants of man."[54]

Marcuse does not go this far, yet he considers Fourier's utopia superior to Marx's precisely because it includes nature.[55]

CHAPTER 3

The Ground of Absolute Freedom in Eros

IN A recent study, "On Herbert Marcuse and the Concept of Psychological Freedom," Myriam Miedzian Malinovich notes that the concern with psychological freedom is primarily a twentieth-century phenomenon. In the eighteenth century, the concern was with social, political, and religious freedom. In the nineteenth century, concern shifted to economic freedom.[1] Psychological freedom, notes Malinovich, has been the concern of three major groups:

1. The existentialist writers, including Sartre and Heidegger, who have been concerned with the quest for authenticity, the shedding of the repressive "ready made" life that society shackles us with in favor of true spontaneity.
2. The psychoanalytic writers, particularly the psychoanalytic left (Reich, Laing, Perls), who have been concerned with liberation from almost all of society's rules, particularly those regulating sexual behavior.
3. The Hegelian Marxists, including Marcuse, who have focused upon the alienation of the worker in capitalist society, and have argued that only in a genuinely socialist society could human beings be free to realize their true nature.[2]

It is the virtue of Malinovich's piece that it recognizes that Marcuse seeks to weave together all three strands of thought on psychological freedom. Thus, it is potentially misleading to refer to *Eros and Civilization* as a synthesis of Marx and Freud, as the book is so frequently characterized. Though the book is certainly this, it also expresses the existentialist concern—which I have characterized in terms of Marcuse's understanding of alienation as alienation from the objective world itself—with the authentic expression of self.[3] Neither the Marxist concern with alienation from capitalist labor nor the Freudian left's concern with erotic freedom adequately characterizes the depth of freedom that Marcuse seeks. Only a perspective that addresses what Marcuse regards as the ultimate shackle on authentic human spontaneity—a natural world that gives up its fruits unwillingly—captures the multi-dimensionality of Marcuse's concept of freedom.

Erotic self-sublimation, in which the pleasure principle would govern itself rather than being governed by the reality principle, is the psychological basis of absolute freedom for Marcuse. Many have argued that Marcuse misinterprets Freud on the process of sublimation. Sublimation could not take place by the overflow of erotic energy to other components of the psyche as Marcuse suggests because sublimation is not the primary process that Marcuse envisions it to be. Sublimation is instead an expression of the mature ego's attempt to bring the individual's needs into harmony with opportunities for their satisfaction. In this way sublimation serves eros.[4] However correct these critics may be, the accuracy of Marcuse's interpretation of Freud is not at issue here. At issue is how Marcuse's vision of freedom requires a world in which human labor is no longer necessary, and what the role of science is in this vision. Should Marcuse be correct and Freud mistaken about the very possibility of erotic self-sublimation, its actual realization would depend upon a world in which machines would do all the work. Marcuse links human freedom and fulfillment to the most ambitious scientific and technological project imaginable: total automation.

The "Dialectic of Civilization"

The "dialectic of civilization," according to Marcuse's interpretation of Freud, works like this. Culture demands the sublimation of the erotic drives so that the psychic energy that would otherwise be directed toward immediate gratification can be inhibited in its aim and rechanneled into work. However, such repression enhances aggression, because the desire for pleasure is frustrated and also because repression leads to guilt regarding desires to transgress social sanctions, and this guilt manifests itself in aggression. The outcome is that erotic impulses that would be able to "bind" aggression by directing potentially aggressive energy toward social tasks are themselves weakened, and so require even higher levels of repression to control agression, which weakens eros still further.

> Culture demands continuous sublimation; it thereby weakens Eros, the builder of culture. And desexualization, by weakening Eros, unbinds the destructive impulses. Civilization is thus threatened by an instinctual de-fusion, in which the death instinct strives to gain ascendancy over the life instincts. Originating in renunciation, . . . civilization tends toward self-destruction.[5]

The influence of the dialectic of Enlightenment upon Marcuse's "dialectic of civilization" is apparent. Indeed, one might well say that the "dialectic of civilization" *is* the dialectic of Enlightenment, expressed in Freudian language. The Freudian perspective does add a new emphasis, however: not merely man's relationship with a sparse nature but the very possibility of civilized relations with other men, depend upon high levels of repression with the consequent sacrifice of erotic aspects of self, of pleasure. Yet, this influence of Freud upon Marcuse should not be taken to mean that Marcuse's is somehow merely a reinterpretation or reworking of Freud. Though Freud does question whether civilization is worth the price, for him there is really no alternative.[6] For Freud, aggression is a primary drive not reducible to lost opportunities for satisfaction, and guilt is an unavoidable consequence of civilization.[7] To call Marcuse's version a "dialectical *Aufhebung*" of Freud, or something like this, would sidestep the issue of how profoundly different their fundamental assumptions are. To make Freud's categories "historical" is really to abandon Freud, a point that is widely recognized.[8] In order to appreciate *Eros*, it is best to see it not as a reworking of Freud but as Marcuse's version of the dialectic of Enlightenment, adapted to an age in which man's final triumph over nature—the abolition of labor—seems possible. With this triumph comes an entirely new historical possibility: that machines could do all the work. The sacrifice of self required by all previous generations to tame nature would thus no longer be necessary.

The psychic roots of the "dialectic of civilization" also reveal its solution, says Marcuse, if we consider that Freud's biological categories have a social component. They are subject to historical modification. If, says Marcuse, the nirvana principle (the desire for absolute peace and cessation of stimulation—the peace of the womb) is the basis of the pleasure principle, then is not the death instinct really an unconscious retreat from pain and want, an expression of the eternal struggle against suffering?[9] This idea is made plausible, says Marcuse, by Freud's last metapsychology, in which the instincts are not defined in terms of their origins or organic functions but in terms of the direction they give the life process. This suggests that the instincts may not be truly dualistic in structure, but instead made up of a displaceable energy which is able to join forces with either the erotic or destructive impulses: "Never before has death been so consistently taken into the essence of life; but never before also has death come so close to Eros."[10]

It is this idea, to jump ahead a bit, that allows Marcuse to argue that now that the primary work of culture has been accomplished, in the sense that in the advanced industrial world there exist enough material resources to satisfy everyone's basic needs, it is in principle possible to relax repression. Under decent conditions a nonrepressed eros would not lead back to nirvana but would itself build culture, though on a more pleasure-oriented basis. This is what Marcuse calls nonrepressive sublimation. Why this appears as such an impossible dream, suggests Marcuse, in light of the high levels of aggression in advanced industrial society, is that a great deal of contemporary aggression stems from the unconscious realization that high levels of repression, once necessary to conquer scarcity, now only serve to perpetuate the unequal distribution of scarcity.

Science and technology, according to Marcuse, "absorb" the modified destructive instincts. In this sense these aggressive drives are put to good social use. However, in this process socially useful erotic energy is more sublimated than aggression, and the balance is tilted in favor of aggression. To be sure, this aggression is diverted toward the external world, but it remains aggression. Furthermore, since this aggression is not adequately "stabilized" by eros, it ultimately accepts no substitutes: not matter, not nature, but the destruction of life itself is the ultimate goal.[11] "Then, through constructive technological destruction, through the constructive violation of nature, the instincts would still operate toward the annihilation of life."[12]

Why are the erotic instincts currently incapable of "binding" the aggression expressed by science and technology? (Though Marcuse employs terms such as the "stabilization" or "binding" of aggression by eros, he does not characterize the psychological processes by which this would occur. These terms serve as "black boxes"; i.e., they characterize a posited process whose mechanism is unexamined.) Repression under high levels of technological progress not only weakens eros but also calls forth even more aggression, as individuals unconsciously recognize that this progress eliminates the need for repression. However, there is an even more fundamental reason. Eros loses its power to bind aggression unless it is free from social control. Eros stems from resistance to the reality principle, from resistance to the demand that pleasure be postponed.[13] This is why, according to Marcuse, the controlled, or repressive, desublimation of eros does not work to bind aggression. Contemporary sexual liberation, which can be seen not only in the relaxation of morals but also in such things as "sexy clothes" or "sexy cars," simply serves the prevailing social order. Real eros rejects such controlled, socially sanctioned pleasures. For eros to be able to penetrate the "dialectic of civilization" and thus stabilize technological aggression, nothing less than a total transformation of society is necessary. For eros to become the builder of culture—for eros to be self-sublimating—it must be absolutely free from external control. Such freedom requires that the expression of eros be completely unhindered. It requires nothing less than the abolition of labor, the ultimate constraint on eros. There is thus an all or nothing quality to the good life. It will not readily be implemented piecemeal, for erotic self-sublimation requires the absolute freedom of eros. This, in turn, requires the virtually complete automation of labor, perhaps the most momentous scientific and technological task ever undertaken.

Under a social order governed by eros, man's alienation from labor would be complete. By this phrase Marcuse means that the automation of labor could so reduce labor time that individuals would no longer need to find satisfaction in their work. They could devote themselves full time to seeking gratification elsewhere. Marcuse expresses it this way: "The more complete the alienation of labor, the greater the potential of freedom: total automation would be the optimum."[14] Nothing has changed since Marcuse's 1933 article on labor, with the exception that since World War II it now seems possible that machines really could do all the work. "The realm of necessity, of labor, is one of unfreedom. . . . Play and display, as principles of civilization, imply not the transformation of labor but its complete subordination to the freely evolving potentialities of man and nature."[15] The idea of erotic work, which Marcuse refers to several times, refers primarily to the social relations of building culture—life in the family and concert hall—and secondarily to the social relations among workers, such as feelings of friendship and solidarity among work groups. It does not refer to the actual task of laboring itself, that is, the relationship of man to the object of labor.[16] It is in this context that Marcuse calls eros a "prop" for "work relations," but not for labor itself.[17] Marcuse does state that it is the purpose and not the content of an activity that marks it as work or play.[18] This seems to suggest that under erotic social relations even such activities as ditch digging could be pleasurable. Yet, ditch digging could be pleasurable only if it were a hobby done entirely for its own sake. The purpose at issue here is only whether the work is necessary. It is the necessity of work that marks it as a constraint on human freedom and thus shows it to be labor.[19] In fact, one could have "derived" this conclusion from the concept of eros itself in connection with Marcuse's absolute concept of freedom.

Marcuse's view of the relationship between labor and freedom is thus quite consistent, in spite of taking a few twists and turns under the influence of eros: they are incompatible. Science and technology serve freedom in only one way: they are concerned with labor so that man need not be. The realm of labor must forever be the realm of unfreedom no matter how it is socially organized, because in labor man must conform to the "law of the thing." The "law of the thing" is nothing less than the objectivity of the natural world itself, its independence from human needs and wants. Science and technology serve absolute freedom by laboriously conforming to the "law of the thing," so that man need not forever do so. Anthony Giddens puts it this way: "Freedom and servitude are not, in Marcuse's theory, phenomena of politics, or even of power more broadly understood. Freedom, Marcuse repeatedly argues, is to be interpreted in relation to satisfaction of need."[20] Complete satisfaction of need, Marcuse tells us, requires the abolition of worldly constraint (i.e., the "law of the thing").

Marcuse is not writing of a world in which men would be laborers in the morning and poets in the afternoon. He is writing of a world in which even the writing of poetry would be a detour from genuine gratification. In this new world perhaps many of those tasks once undertaken as labor would be once again undertaken as play. The actual content of the activity might remain, but its being self-imposed, much as individuals impose challenges upon themselves when they play a game, would mark it as an act of freedom. Yet, Marcuse has defined freedom and pleasure so radically—not only are they incompatible with necessity but also with worldly objectivity—that virtually any activity that impinges on the free play of eros is incompatible with them. Even a life of hobbies done for their own sake, such as butterfly collecting, imposes objective demands upon individuals. Such objective demands stand terribly close to the "law of the thing," whether or not the demands are freely accepted by individuals. For it is nature's rules, not man's, that set stringent limits as to how the hobby may be practiced or the game played. The butterfly does not rush to meet the collector's net, and the ball does not roll uphill unaided.

Unless man is to unfold his sensual self in complete isolation from the world, surely a sterile ideal, even in play he is faced with the demands of worldly objectivity. Not merely a lack of free time but the objectivity of the natural world stands as a barrier to absolute freedom. Thus, Marcuse sometimes suggests that once man has technologically mastered the natural world and revolutionized the social relations of labor, the natural world would itself change. Marcuse is vague about the details, but the principle is clear. The world would no longer be such that men and women would be compelled to conform to its demands. Ultimately this means discovering that it is man who makes and enforces ontological (man-world) distinctions. "Such a world could (in a literal sense!) embody, incorporate, the human faculties and desires to such an extent that they appear as part of the objective determinism of nature."[21] This is indeed the transformation (i.e., elimination) of all those "neo-Kantian" distinctions between subject and object that Marcuse found so alienating in the *Künstlerroman, Hegels Ontologie*, and his article on Dilthey. Then, this elimination was the task of art and philosophy. In his later work it has become the task of science and technology. How this line of thought might lead to a "romantic" new science is apparent. Yet, appearances can be deceptive. The new science is not based merely upon the incorporation of nature; it also seeks guidance from nature (this latter aspect of the new science is discussed in the next chapter).

Eros and Civilization elevates an instrumental interpretation of science into a vehicle of absolute freedom. It also contains a theory of how man would overcome—incorporate—the separate existence of the natural world, and so transform it into an expression of human freedom. Indeed, what is so striking about Marcuse's project is how readily these two themes coexist. In a world governed entirely by self-sublimated eros, says Marcuse, the essence of being would itself be erotic, for the play of eros and death corresponds to the traditional metaphysical play of being and nonbeing.[22]* With this assertion Marcuse draws the source of man-world distinctions not merely into human history, as in the essay on Dilthey, but into human nature. Indeed, Marcuse must struggle to make eros, which applies only to people, a universal principle of being. He does so by arguing that because inorganic matter is the end or telos of the death instinct it too is linked to human life.[23] If a new science could discover a new nature by remaking nature in man's own erotic image, then nature too would become, it appears, virtually an expression of eros. Nature would become not only—but merely—a social category. This interpretation helps place in context Marcuse's occasional suggestion that the liberation of nature would let nature be what it would like to be.[24] It is apparent that an erotic orientation toward nature actually has little to do with "letting" nature be anything. Rather, it is an expression of the desire to make nature an extension of man's own erotic impulses. History, says Marcuse, "is the negation of nature. What is only natural is overcome and recreated by the power of Reason."[25]

Marcuse considers the claim that science is rational enlightenment about the world a regressive claim, because it forecloses possi-

*It is difficult to see how this scheme could be applied meaningfully to nonhuman organic matter: plants and animals. Do they too partake of eros? Furthermore, only Freud's scheme, which Marcuse wishes to overthrow, links eros to inorganic matter. Marcuse argues that the death instinct is socially conditioned, and thus contingent. Under nonrepressive conditions the neutral displaceable psychic energy once channeled into aggression would be rechanneled into eros. Under nonrepressive conditions there would thus no longer be an "ontological link" between inorganic matter and eros. Marcuse's speculations do not apply to his own scheme.

bilities. To claim to know the way things truly are obviates the potentiality of things: how they might become. This is the perspective of critical negativism. "Traditional science was in fact more subject to the existent than was great philosophy. It was not in science but in philosophy that traditional theory developed concepts oriented to the potentialities of man lying beyond his factual status."[26] In this respect Marcuse is heir to the critique of the dialectic of Enlightenment. At the same time, Marcuse makes science and technology the instruments of absolute freedom. The difference with most instrumental interpretations is that Marcuse's is not merely an instrumental view of scientific theory or knowledge. For Marcuse science is an instrument for the realization of man's essence. It removes the final natural barrier to human self-realization: the time spent in labor. Denying science the right to know reality while granting science enormous powers, more enormous even than the powers granted science by the classical utopian theorists such as Bacon and More—for whom science and technology "merely" ease the conditions of man's real existence rather than being vehicles for the realization of his essence[27]—must lead to contradictions. In Marcuse's case the instrument that can liberate mankind loses any authority to speak to the empirical limits of human freedom.

A Narcissistic Strategy?

A recent argument by Christopher Lasch, in *The Minimal Self*, suggests that even could the contradictions in Marcuse's strategy discussed above be resolved, his project would remain fundamentally flawed. Lasch argues that Freud's later work, as well as the work of Melanie Klein and her followers, the object-relations theorists, demonstrate that the thesis that repression originates in the subjection of the pleasure principle to the reality principle is too simple.[28] Marcuse relies on outdated psychological theory. Instead, says Lasch, the problem of scarcity is better understood as being rooted in separation anxiety, and, ultimately, the fear of death. In this respect, says Lasch, Norman O. Brown, in *Life Against Death*, was closer to the mark than Marcuse.

> For Brown, the "lack of sufficient means and resources" derives not from the social organization of production but from

> the very urgency of instinctual demands. "Scarcity" is experienced first of all as a shortage of undivided mother-love.[29]

Because the primal scarcity is undivided mother-love, not the scarcity of the larger natural world that requires labor, even the total automation of labor would not lead to unadulterated peace and joy. The total automation of labor, even were it possible, would only modify the developmentally later subordination of the pleasure principle to the reality principle. It would do little to address the fear of death, which is the adult correlate of separation anxiety. At a deep psychological level the separation from mother, from nature (especially as this separation takes the form of scarcity), and ultimately from life itself, coincide, and build upon each other.

Not only does Marcuse's strategy of total automation address only one aspect of this sedimented complex, suggests Lasch, but it does so in a regressive, narcissistic fashion. Lasch interprets secondary (pathological) narcissism as the failure to come to terms with the psychic tension that results from the desire for union when confronted with the fact of separation. Narcissistic strategies seek, in various ways, to mitigate this tension by fantasies of unity or by fantasies of complete self-sufficiency, by the denial of any need for external objects (the external world) at all.[30] It is from this perspective that Lasch sees Marcuse's ideal of total automation as the ultimate narcissistic denial of humanity's dependence upon the natural world. However, suggests Lasch, such denial is hardly unique to Marcuse. It is at the root of humanity's attempt to dominate and control nature. Indeed, one can see this aspect of Lasch's interpretation of narcissism as a psychoanalytic version of the critique of instrumental reason, an interpretation Lasch seems to encourage.[31]

Lasch's analysis is insightful. Indeed, in Marcuse's preference for Narcissus over Odysseus and Prometheus as a "culture hero," one can perhaps detect a wish to deny the reality of death itself.[32] Narcissism rejects separation. "The point of the story is not that Narcissus falls in love with himself, but, since he fails to recognize his own reflection, that he lacks any conception of the difference between himself and his surroundings."[33] In rejecting separation, narcissism rejects the very possibility of death, insofar as it rejects the fact of separation that must precede death. It is separation—from mother, from world—that makes death a conceptual and psychological possibility. In fact, several of Marcuse's comments in

Eros to the effect that death is "perhaps even an ultimate necessity" (i.e., perhaps it is not!) and that death should not be perverted from a biological fact to an ontological essence suggest that Marcuse occasionally hopes that humanity's triumph over nature is to include triumph over even the ultimate natural constraint.[34] This would indeed be narcissistic fantasy. Yet, as is frequently the case, Marcuse's strategy is more subtle and complex than his critics recognize. As was suggested previously (and as will be demonstrated in the next chapter), Marcuse uses several strategies to come to terms with the separateness of nature, its indifference to humanity's needs and wants. While the strategy of overcoming nature via science and technology dominates, the strategy of the ontological incorporation of nature, as well as the strategy of seeking guidance from nature, are also present. Indeed, it is the latter strategy that best expresses the new science.

To be sure, these strategies do not always fit well together. There is an element of mutual contradiction, as will be shown. Nevertheless, it would be a misreading of Marcuse's project to see in these strategies simply the poles of the narcissistic denial of separation: the fantasies of unity and utter autonomy. Indeed, Lasch's own solution to the problem of narcissism suggests why such a reading of Marcuse's project is mistaken. There are, says Lasch (in what is surely a contentious claim), no straightforward and free activities. Virtually all of life, including such apparently free activities as art, play, and lovemaking, is an attempt to come to terms with the anxiety and pain of separation. The appropriate response to the reality of separation, says Lasch, "appears to lie in a creative tension between separation and union, individuation and dependence."[35] Such a response accepts the omnipresent reality of separation that it seeks to come to terms with.

Whether Lasch is entirely correct or not (e.g., whether or not all of life is drawn into the problem and pain of separation) is not at issue here. One could, though Lasch does not, draw politically quietistic conclusions from such a viewpoint. What is at issue is that Marcuse's three strategies for dealing with the separateness of the natural world seem to reflect something of the creative tension that Lasch recommends. Only a reading of Marcuse that reduces his program to one strategy, that of overcoming nature with science and technology, could see it as merely an expression of narcissistic denial. Marcuse's program is that too, of course. However, as Lasch suggests, any adequate solution will involve this element. The question is rather whether there are other elements that modify and constrain such narcissistic denial. Though these elements in Marcuse's work are often not so effective as they might be, it will be shown that they are nonetheless present. If Marcuse does not strike quite the right balance, there are nonetheless balancing elements in his project that Lasch neglects.

CHAPTER 4

Marcuse's New Science and its Dissolution in Freedom

ONLY IN *One-Dimensional Man* (1964) does Marcuse elaborate upon his criticism of science. However, the foundation of this criticism can be found in each of the works previously examined. Barry Kātz states that no new theoretical material is introduced in *One-Dimensional Man*. It is a compilation and weaving together of earlier themes.[1] In general, this is correct, even for Marcuse's discussion of science. There is, however, an exception to this generalization. One sees in *One-Dimensional Man* an aspect of the new science not seen before: a new science that would let itself be guided by nature, rather than one that seeks to incorporate nature into human eros. However, the foundation of even this aspect of the new science can be found in Marcuse's earlier works, especially "On Hedonism" (1938) and *Eros*. In the end, however, this aspect of the new science, even when coupled with its counterpart (that aspect of the new science that seeks to incorporate nature into human eros), gives way before an instrumental science that could realize human freedom by overcoming nature.

There are three stages to Marcuse's indictment of the modern science: (1) science is inherently instrumental; (2) this makes science dependent on external ends, for as an instrument science can only serve goals set for it by society; and (3) this is the basis of the irrationality of science, in which it fails to take the responsibility of goal setting into itself. "Pure science has an inherently instrumental character prior to all specific application; the Logos of pure science is technology and is thus essentially dependent on external ends. This introduces the irrational *into* science, and science cannot overcome *its* irrationality as long as it remains hidden from science."[2] Marcuse also mentions a fourth stage, in which science is not only inherently technology, but inherently a particular technology: the technology of social control and domination. That is, Marcuse sometimes seeks "to demonstrate the *internal* instrumentalist character of this scientific rationality, by virtue of which it is *a priori* technology, and the *a priori* of a *specific* technology—namely, technology as a form of social organization and control."[3]

Marcuse develops this fourth stage less extensively than the other three. The fourth stage is perhaps best seen as a legacy of the dialectic of Enlightenment: to dominate nature requires such a sacrifice of the self's desire for peace and satisfaction that the self becomes especially vulnerable to manipulation. This manipulation may take the form of the clever demagogue's exploitation of the self's desire to return to nature, in which the demagogue would channel this desire in regressive directions, such as National Socialist "volkish" ideology of "blood and soil."[4] Or, the self may be manipulated by something more abstract: an industrial system that promises to meet every conceivable material and psycho-sexual need. The price of such satisfaction is conformity, as every alternative to this system is made to appear by the power of the system itself as a regression beyond the pale of rationality. It is, of course, this latter form of manipulation that concerns Marcuse in *One-Dimensional Man*. However, the psychological prerequisites of such manipulation are more thoroughly analyzed in Marcuse's "The Obsolescence of the Freudian Concept of Man" (1963).[5]

Another way to read this fourth stage, one which in my opinion does not pay sufficient attention to it roots in the dialectic of Enlightenment, is to see it as an expression of Marcuse's tendency to explain too much by means of the concept of the "technological totality" (i.e., the political-social-economic-military-industrial complex). Though Marcuse intends, this criticism would read, to link his critique of instrumental science to his social and political analysis—in order to show that science is the instrument of a particular social and political order—his conjunction of science and politics is in fact too vague and general. Too often Marcuse's argument comes to depend upon the impressionistic and metaphorical use of language, such as his concept of the "technification of domination."[6] Yet, while this criticism of Marcuse's sometimes vague use of the "technological totality" as an explanatory concept seems warranted in many cases, the fourth stage cannot be dismissed quite so readily.* Though Marcuse never satisfactorily develops his argument regarding the fourth stage, its origins in the dialectic of Enlightenment and not merely in his more recent "technological totality" are apparent, as will be shown.

Marcuse sees formal logic as the first and decisive step toward scientific thought. The accomplishment of formal logic, says Marcuse, was to make the mind susceptible to abstract generalization, so that contradiction would lose its place as the fundamental category of existence.[7] Formal logic accomplishes this task, says Marcuse, by separating is and ought. The former category comes to be seen as part of the existence of things themselves, whereas questions of ought become the subject of disciplines concerned with right knowing (epistemology) or right action (ethics). In this way conflicts between essence and appearance are minimized. Not only are matters of essence and appearance made the subject of separate disciplines but this separation itself justifies the idea that there exist two distinct realms of experience, signified by the polarities is and ought, potentiality and actuality, change and permanence. Dialectical logic (similar to what I have called critical negativism), on the other

*The criticism is familiar. Marcuse, it is argued, tends to mix scientific, technological, political, economic, and cultural trends in such a way that even if one accepts the thesis that these trends function together to preserve the existing order it is immensely difficult to determine their relationship. One outcome of this "totalistic" tendency is that the overwhelming tightness of repression, its intrusion into every aspect of life, is overemphasized. Another is that Marcuse's model tends to be deterministic and mechanistic insofar as it presumes that the level of technological development has itself become the dominant variable in social change. It is from this perspective that Claus Offe, in "Technik und Eindimensionalität. Eine Version der Technokratiethese?", argues that Marcuse believes technology to be a more important source of repression than politics or economics, pp. 73–88. Alasdair MacIntyre's *Herbert Marcuse* is a classic criticism of Marcuse's vague and holistic criticism, esp. pp. 99–106. See too Wolfgang Fritz Haug, "Das Ganz und das ganz Andere. Zur Kritik der reinen revolutionären Transzendenz."

hand, understands the relationship between is and ought as an ontological one inherent in the structure of being itself. Dialectical logic thus does not seek to analyze reality independently of the distance between is and ought. Indeed, dialectical logic knows this distance, and the tension in reality that it represents, to be the most real thing of all. Formal logic, precisely because it does not regard the copula "is" (as in all *x* is *y*) as also containing a judgment over the rightness of existence, acts—however unconsciously—to stabilize the existent, because it separates analysis and criticism. The absolutely fundamental character of Marcuse's criticism is apparent here, a legacy of the dialectic of Enlightenment. In this view Aristotle must be one of the first exponents of technical reason.[8]

If formal logic is the first step by which nature comes to be seen as an object of calculation and control, the instrumentalism of science is the culmination of this tendency. One must ask, however, what it really means to say that science is instrumental. After all, it has been shown previously that this term lets itself be used in a variety of contexts. At one level Marcuse seems to mean simply that scientific knowledge lets itself be used for any purpose, good or evil. There is nothing about the knowledge of the way things are that prevents this knowledge from being misused. In this sense scientific knowledge is instrumental because it can be used as a tool to reach various ends.[9] Marcuse means more than this, however. For Marcuse, the instrumentalism of science rests on two additional theses: science uses numbers to represent natural objects, and science rests in the empirical world.

The numbers thesis suggests that science experiences nature only in terms of how it may be calculated. However, as Aron Gurwitsch asks, what does it mean to say, as Marcuse so often does, that science calculates nature?[10] Calculation is an abstract mental process, particularly in its more developed forms, such as formalization and algorithmization. As such, calculation pertains directly to theories and only indirectly to nature. The relationship between calculation and nature is mediated by prediction. The distinction between saying "science calculates nature" and "science calculates theories about nature that are then tested against nature" is significant. While the former statement suggests that scientists look at nature as though it were nothing more than numbers waiting to be manipulated, the latter interpretation suggests that the goal scientists project onto nature—for science is of course not value-free in this respect—is to understand nature, to explain it.[11] It is this goal that numbers serve as an instrument. This consideration of the function of numbers in scientific explanation does not, of course, counter all arguments that might be employed to demonstrate the instrumentalism of science. Habermas' charge of instrumentalism rests upon different considerations, as will be shown. Nevertheless, Gurwitsch's distinction does point out that the use of numbers in scientific explanation is not automatically equivalent to the "calculation of nature." Indeed, it suggests that the image of scientists "calculating nature" is just as metaphorical as the Baconian image of scientists "stalking nature" or "torturing nature" in order to capture her secrets. How revealing such metaphors are about scientific explanation is questionable.

The second thesis employed by Marcuse to demonstrate the instrumentalism of science is that science rests in the empirical world and thus is not transcendent. Whether the instrumentalism of science is even implied by its nontranscendence is certainly open to question. What Marcuse seems to mean, however, is that science—following in the footsteps of formal logic—separates is and ought, actuality and potentiality. It does so by isolating empirical phenomena within a particular slice of history: the immediate present. Marcuse states that in reality is and ought, actuality and potentiality, always occur together. This idea can be explained as follows. Let us employ a historical perspective, rather than the empirical present, as our point of view. From the historical perspective it can be seen that the way things once were is no longer the way they are now. Things change and will presumably continue to do so. In this process of change the potentiality of things becomes their actuality. The way things could develop is often (though of course not always) the way they do develop. It is this process of change and development that is reality. The empirical focus upon the immediacy of things only captures a thin slice of this process by which things become what they are: the present. It is the empirical focus upon actuality and immediacy that is an abstraction from reality. While formal logic has caused us to forget that it is the separation of potentiality and actuality that is the real abstraction from reality, science has reinforced this illusion through which we confuse the empirical present with reality. Science has reinforced this illusion precisely because it has been such a powerful and useful one. Operating under it men and women have learned to transform the physical face of the earth.

Marcuse argues that modern science is not merely instrumental but also, and quite paradoxically, idealistic. The theories of science conform to an abstract mental picture of the world rather than to the world itself, which is characterized by the virtual inseparability of is and ought. In fact, it is the combination of instrumentalism and idealism that is particularly deleterious. Science tends toward idealism, says Marcuse, because its emphasis on theory causes the objective world to become more and more dependent upon men for its constitution.[12] As matter in the world becomes comprehensible in mathematical equations, which translated into technology transforms this matter in a useful fashion, matter itself tends to lose its status as independent substance. The source of this claim appears to be Marcuse's interpretation of Husserl to the effect that science has forgotten that its abstractions refer back to the lifeworld.[13]*

Scientific instrumentalism, suggests Marcuse, is inherent in logical-mathematical abstraction, which regards nature as fungible units for the purposes of calculation.[14] However, instrumentalism is enhanced by idealism. If nature has no essence itself, it becomes merely the field of man's experimental operations. Marcuse regards operationalism, in which concepts are defined by how matter is manipulated, as the most extreme form of this tendency, though he sees it in all scientific endeavor. An empty nature, combined with a logical-mathematical perspective, leads to the idea that we can only understand nature by manipulating it: first under experimental conditions, then with technologies. Again, this tendency is particularly intractable because the success of technological manipulation seems to confirm the truth of the instrumental approach.

In this context, it might be argued that the work of Husserl, whose lectures Marcuse attended at Freiburg, exerts a strong influence upon Marcuse's conception of science.[15] This would be seen not merely in Husserl's specific claim that science has lost its roots in the lifeworld,[16] but also in the attraction Husserl's overall philosophy might hold for Marcuse. For Husserl's philosophy is "con-

*It is disconcerting to find Marcuse admitting that he has difficulty finding examples of the idealism of science in the writings of the philosophers of science. See *One-Dimensional Man*, pp. 149–53. For example, Marcuse argues that Karl Popper is "struggling with an idealistic element" because Popper believes that while better theories replace worse ones, we may never know for certain if a theory is true (*One-Dimensional Man*, pp. 151–52). However, to use such a position as evidence for the idealism of science it seems that one must collapse any distinction between the truth of theories and the world they purport to be about.

crete": it seeks to apprehend the things themselves and in so doing avoids positing an abstract, idealized, and unknowable "nature in itself."[17] However, while Husserl surely exerts some influence upon Marcuse's understanding of science, to argue that the influence is dominant requires that one assume that Marcuse is unaware of, or at least chooses to ignore, quite contradictory interpretations of Husserl's work. For instance, Aron Gurwitsch argues that Husserl assumed that science only mediately rests in the lifeworld. The existence of the lifeworld itself presupposes an active, constitutive consciousness.[18] In any case, it is unnecessary to assume that Marcuse bases his new science upon Husserl's work to any significant degree. What Marcuse rather seems to do (it will be shown) is to adopt the assumption of Horkheimer and Adorno that nature is a value in itself; i.e., ethical value inheres in aspects of nature itself. The alternative, Marcuse suggests, is a view of nature that sees it as so irrelevant to human purposes that it becomes a pure field for human self-assertion. Yet, we have seen that Marcuse risks his own version of this tendency: the projection of human erotic drives onto nature. (Marcuse would presumably respond that these drives are not merely projected onto nature; rather, they have their home in nature, including human nature, which serves to link human eros with the eros of the natural world.)

While the influence of Husserl's "concrete" philosophy upon Marcuse's analysis of science is debatable, it is undeniable that Marcuse's more recent analysis contradicts his 1928 analysis of science in "Contributions." Then, Marcuse asserted that science rests in the empirical world. Furthermore, it should rest in the empirical world; only the sciences of man are genuinely historical and hence legitimately dialectical. Then Marcuse accepted a realistic interpretation of science. However, in *One-Dimensional Man* Marcuse rejects both these assertions. Now science rests in the minds of scientists at least as much as it does in the empirical world. For this reason Marcuse indicts science as idealism; at the same time, he claims that science is not idealistic enough. It mistakenly believes reality is given and thus does not attempt to formulate its concepts and theories in a way that points beyond this reality. This contradictory indictment of science is an almost pure expression of the dialectic of Enlightenment, which argues that what is missing in modern thought is any standpoint between empiricism and idealism: nature becomes mere stuff, whereas values are elevated to the abstract realm of ideas, where they become irrelevant. Marcuse's solution is apparent: to locate these otherwise impotent values in nature, and thus bring them back to earth, to make them real—indeed, empirical—once more. Yet, this is a solution to which Marcuse devotes more rhetoric than serious analysis. Absolute freedom remains the paramount goal, and absolute freedom from the objectivity of nature is best realized, Marcuse appears to conclude, by an instrumental science.

Though Marcuse ultimately accepts the instrumentalism of modern science, he does not do so willingly. Indeed, he seeks to demonstrate that the instrumentalism of contemporary science allows irrational social goals to seep into science in such a way that science itself becomes irrational. Presumably in a fully rational social world the permeability of instrumental science to social values would automatically cease to be a problem. There would no longer be any distinction between scientific and social rationality; there would no longer be anything bad to seep in. Today, however, science is not yet coincident with reason, which grasps the inseparability of is and ought, potentiality and actuality. This state of affairs indicts science, not just society. "In as much as Reason remains non-manifest in science, scientific rationality is not yet the full rationality of science," states Marcuse.[19] Science is indictable for what it is *not*: historical and dialectical in the manner of "Reason" with a capital "R."

> For the *scientific* subversion of the immediate experience which establishes the truth of science as against that of immediate experience does not develop the concepts which carry in themselves the protest and the refusal. The new scientific truth which they oppose to the accepted one does not contain in itself the judgment that condemns the established reality.[20]

It is this judgment of the established reality, in which ought confronts is, that seems to be the essence of reason for Marcuse.

Yet, the way in which Marcuse formulates this claim suggests another dimension to it as well. The truths of history and ethics, which science ignores, reside not just in a richer philosophical tradition concerned with reason but in nature (including human erotic nature) itself, Marcuse seems to be saying. This is why the idealism of science is so deleterious. It empties nature of its inherent meaning, its inherent rationality. While Marcuse criticizes science for its nontranscendence, it is he who wishes to return science to the lifeworld, albeit on the basis of a different—sensuous—experience of it. Science rests in the wrong empirical world: the world of the purely present. Science is transcendent, but it transcends the wrong things. It transcends an order and meaning inherent in nature itself. The way in which Marcuse proposes to return science to the right lifeworld is discussed shortly.

Recall the shift, in the quotes used at the beginning of this chapter, from "into" to "its" (p. 50). It would seem that the instrumentalism of science, which allows society's irrationality to be injected "into" science, as science lets itself be used for virtually any purpose, would make science merely an arational enterprise. In fact, Marcuse pursues this point in "Industrialization and Capitalism in the Work of Max Weber," as well as in "Some Social Implications of Modern Technology" (Marcuse also clearly states the "science is a technique of domination in itself" position in the conclusion of "Industrialization and Capitalism," but he does not pursue it there).[21] Yet, Marcuse's criticism is more radically based. Science itself is irrational, for it does violence via its abstractive method to an essentially rational order inherent in the world. Here is Marcuse's argument, albeit not a developed one, for the claim that science is inherently an instrument of domination and control. Science overwhelms the natural rationality of the world, substituting for it one that responds to the narrower instrumental rationality of humankind. Yet, this radical critique rests only indirectly upon the internal instrumentalism thesis. It rests directly upon the "idealism" of science, which empties nature of this inherent order. This is why returning science to the lifeworld is so important to Marcuse.

The New Science

In order to show that science could rest in another lifeworld, Marcuse rejects what he sees as Jean Piaget's argument that physics, which leads back "necessarily" to logical and mathematical abstraction, is rooted in the physical structure of thought. Habermas' rejection of Marcuse's new science, it will be shown, is in large measure based upon Habermas' acceptance of Piaget's argument. Instead, argues Marcuse, Husserl has shown that logical and mathematical abstraction rests on a way of "seeing" the world that arose in a particular historical context, brought about by the need to coordinate empirical regularities revealed by surveying with the more precise ideational regularities required by geometry.[22] By recalling this basis, says Marcuse, the issue is raised whether a new way of seeing the world might not lead to qualitatively new relations between man and nature.

In order to document the possibility of this new way of seeing, Marcuse turns to the "convergence thesis," which projects the convergence of technology and art, work and play. No new theoretical material is introduced to support this thesis. Its basis is found in *Eros and Civilization* and in "On Hedonism" (1939). In "On Hedonism," Marcuse tries to show that happiness can function as a basic goal of men and women. In order that it could, says Marcuse, it must be shown that happiness can be more than a subjective experience. The subjectivity of happiness is why traditional philosophy rejected it as a basic goal. (Aristotle is hardly an exception, it should be noted. Aristotle rationalized happiness, making it equivalent to a successful life of moderation in all things, in order to posit it as the human telos.)[23] One could not derive a general principle of reason from happiness, says Marcuse, because happiness seemed to be an outcome of chance. Only a revolutionary materialism can reconcile reason and happiness, suggests Marcuse, for while it makes sense to talk of a general reason separate from individual happiness, it does not make sense to talk of a general happiness separate from individual happiness. Hedonism is therefore a potentially revolutionary principle. If it is pursued consistently it cannot be an outcome of individualistic competition. Such competition leads back to the reconciliation of individual happiness and general unhappiness, a situation that exhibits the separation of reason and happiness rather than their integration.[24] The problem is how to civilize hedonism and thus make it rational.

The most progressive aspect of hedonism, says Marcuse, resides in its perception of things in an immediate and receptive manner. Marcuse calls this sensuousness.[25] Sensuous perception grasps things as they truly are by penetrating the layers of history that surround them in order to capture their one true form.[26] Sensuous perception is passive and receptive. It lets itself be guided by the object, whereas even dialectics retains an element of the instrumental, insofar as it grasps an object solely under the horizion of human purposes.[27] Dialectics understands a thing by uncovering the process by which it has become what it is: its unfolding in the course of human history. Sensuous perception is ahistorical; it lets itself be guided by the thing, not by human purposes. Sensuousness grasps that aspect of a thing that is historically invariant. This invariant aspect of a thing is by definition its essence.

If, argues Marcuse, pleasure can itself be seen as a type of perception (sensuous perception), then pleasure has an epistemological function. On the basis of its intentional character—it seeks pleasure—pleasure distinguishes between truly pleasurable and unpleasurable objects. In a nonrepressive world, says Marcuse, the senses could take over the task of moral judgment. Good and evil, beautiful and ugly, would be differentiated on the basis of that which gratifies and that which does not.[28] The vast variety of taste and predilection, says Marcuse, would be shown to be the unfolding of a single original sensation of pleasure and its opposite.[29]* In this vein Marcuse examines Plato's study of how pleasure is related to other cognitive faculties. For instance, consider how expectation and memory affect pleasure.† Marcuse concludes that we may draw into the concept of pleasure attributes of cognition having to do with morality and conclude that good men have true pleasures and bad men false ones.[30] In this way true happiness, pleasure, is bound to reason, for it has an inner connection to truth and thus to morality.[31] This is the civilization of hedonism and the ground of the new science.

As an expression of sensuous perception, the new science is thus not dialectical in the manner of, for example, Engel's "dialectics of nature."[32] The new science does not seek to draw nature under the horizon of human purposes, a strategy that is otherwise so dominant in Marcuse's work. The new science thus does not seek to incorporate nature into human categories. In a sense it seeks the opposite. Nor does Marcuse's new science seek to materially overcome nature in the manner of ordinary science. Indeed, by contrast with the new science both these approaches are instrumental, insofar as both view nature from a perspective in which human needs come first. The new science is a special category of knowledge, one that comes closest to expressing the goal of reconciliation with nature. It

*Actually, this scheme would not transform the problem of moral choice, but would merely shift its locus: from consciousness to instinct. Were an act questioned under this assumption, debate would center upon whether act "x" is a genuine instance of instinctual moral sensibility or merely its historical manipulation. What is lost in this theory is the concept of individual responsibility for one's choices.

†If as Marcuse quite reasonably suggests, memory and expectation are a part of pleasure, then how could pleasure also act as an independent, virtually instinctual, judge of good and bad? That is, how could pleasure stand outside of those experiences which give memory its content and give expectation its presumptions regarding the future?

responds to nature, in the hope that by so doing nature will reveal a softer and more pleasant aspect to humanity. The new science is, however, readily overwhelmed by Marcuse's paramount goal, which is hardly compatible with it—absolute freedom from nature.

Under the new science, says Marcuse, all knowledge would itself become an immediate source of pleasure. "Knowledge will no longer disturb pleasure. Perhaps it can even become pleasure," states Marcuse.[33] Knowledge would be bound to pleasure not because reason would itself be experienced as pleasurable—Marcuse considers this to be an unwarranted rationalization of hedonism by Plato and Aristotle—but because sensuousness would itself become rational in the aforementioned sense. Reason is purely a means for Marcuse, not an end. The end is not pleasure in thinking, but pleasure per se. Marcuse suggests that a sensuous attitude toward nature would reveal previously undisclosed aspects of nature, which would require a new scientific body of knowledge to take them into account.[34] These aspects would be the friendly, pleasurable, helpful—in a word, fraternal—qualities of nature. Such an orientation toward nature differs somewhat from the epistemological strategy of incorporation. Incorporation, it will be recalled, seeks to make external nature an expression of the potentially erotic organization of the person, whereas the new science seeks to discover in external nature principles to inspire and guide human action. Such a doctrine is technically known as ethical naturalism. The new science takes its direction from external nature, epistemological incorporation from human nature. Each would eventuate in a virtually identical nature, however, because it is the principle of eros, whether found in world nature or human nature, that would prevail in such a reformed nature. Indeed, this is what establishes the continuity between these strategies, a continuity that Marcuse seems to assume rather than explicitly spell out. Indeed, he does not really distinguish between these two strategies, perhaps because he regards them as continuous. However, while these two strategies may be continuous, how sharply they differ from the strategy of letting an instrumental science and technology take over man's "metabolism" with nature, so that man can be free to unfold his sensual self liberated from the demands of an objective world, will be emphasized.

Marcuse never develops his new science any further. He states, to be sure, that under the reign of the new science causality would be replaced by freedom, and realism by surrealism.[35] Yet, as an instance of the "gaya scienza" Marcuse refers to the liberation of the imagination and playfulness of scientists from the demands of a political and economic system that fosters and rewards only those lines of research that contribute to the hegemony of the modern industrial state.[36] Such an argument is perhaps not conventional, but it is hardly radical either. It is frequently held, for example, that the until recently relatively "overdeveloped" status of physics vis-à-vis biology is in significant measure attributable to the practical relevance of the former in the service of the industrial revolution and warfare.[37] It is interesting but hardly revolutionary to point out that public policy—how much money is spent on what lines of research—will affect the corpus of scientific theory: what is discovered and what is not; what is considered fact and what fancy. Only in the very long run, if at all, do all research paths lead in the same direction.[38]

The claims Marcuse makes on behalf of "sensuous perception" are truly radical. Yet, the apparent realization of sensuous perception, the new science, is discussed in terms that owe more to Marcuse's study of the pervasive influence of the "warfare state" than they do to his epistemological radicalism.[39] Surely a large part of the reason that Marcuse cannot grant the new science a content consonant with its radical foundations is simply that the theory of sensuous perception is so improbable that its realization is virtually unimaginable.[40] Yet, there is another reason as well, which is even more revealing of Marcuse's project. Alienation, for Marcuse, is more than a category of capitalist economics. It refers to the fact that man can never be at home in a world he has not made. Under socialist revolution man may overcome his alienation from *labor* by reappropriating the product of his labor, governing himself democratically in the workplace, and the like. In order to overcome his alienation from a world he finds himself "thrown into," "ontological revolution," as it were (i.e., the incorporation of external nature) is necessary. It is as though for Marcuse not just the table but also the tree from which it is made must be constantly reappropriated in order to overcome man's alienation from a natural world in which he is never at home, but only a laborer. This is seen especially clearly in his 1933 article on the concept of labor.

The Heideggerian influence here is apparent. However, to put it this way is really too academic. Rather, the theme is one that runs through the *Künstlerroman, Hegels Ontologie*, and Marcuse's appropriation of the dialectic of Enlightenment in his "dialectic of civilization." Man experiences the world as an alien place, unfriendly to his deepest longings for peace and joy. His survival depends upon his repression of these inner needs in favor of an instrumental, indeed domineering, attitude toward nature and man. But, the very power of man's repressed desire for unity, particularly with his own erotic nature (a desire that could be traced by an underground history of philosophy beginning perhaps with Aristophanes' praise of the unifying power of love in Plato's *Symposium*) testifies to an original unity. The original unity is what Marcuse in *Hegels Ontologie* calls an "originary synthesis (!) which is the 'measure' of all comparison and all opposition."[41] Until now this "originary synthesis" [*ursprüngliche Synthesis*] has been the stuff of utopian dreams. However, we live in a world-historically new situation, Marcuse believes, in which the end of the necessity of labor is in sight. In this context, the new science is best understood as the scientific and technological potential to overcome the sparseness of nature, developed to such a degree that man would no longer need to repress his own erotic nature, because he would no longer need to deny himself in order to labor. Bound and infused by such an erotic nature, a new science would no longer see the world as an object of domination—even though the potential exists to dominate it utterly—but as a helpmate. The world would become an expression of human erotic powers.

> Such a world could (in a literal sense!) embody, incorporate, the human faculties and desires to such an extent that they appear as part of the objective determinism of nature.[42]

However, one might ask, does not a touch of aggression, or hubris, as Marcuse puts it, characterize this belief that nature would come to reflect human needs? Yes, Marcuse responds, but a touch of hubris in this case is perhaps unavoidable.[43]

How radically, or literally, to take Marcuse's new science is not completely answered by these considerations. Whether E. F. Schumacher or Charles Fourier comes closer to being the paradigm remains uncertain. However, there is an utter radicalism to Marcuse's thought that should not be underestimated. This is revealed, for example, in his preference for such unpolitical thinkers as Heidegger and such abstract philosophical thinkers as Hegel over Sartre. Marcuse prefers Heidegger and Hegel because they do not compromise their philosophy in any way to a bad existence. They epitomize critical negativism, a viewpoint, it was previously pointed out, that has its origins in Marcuse's *Künstlerroman*. Heidegger and Hegel epitomize this attitude because their standards and analysis remain pure, even at the cost of relevance. Sartre, on the other hand, "revives Hegel's formula for the free and rational condition of man. To Hegel, however, the realization of this condition is only the goal and end of the entire historical process. Sartre takes the ontological shortcut and transforms the process into the metaphysical condition of the 'pour-soi.'"[44] For similar reasons, Marcuse considers the bourgeois novel far more radical than socialist realism. The stylized form of bourgeois art holds out a "promise of happiness" that transcends material abundance, worker self-governance, and nonalienated labor.[45] It might be useful to think about the intent of Marcuse's new science in this fashion. It reflects the promise of an "originary synthesis" that can serve as a "measure" [*Massstab*] of the actual contribution of science and technology to human freedom. In this light, ecology, environmental consciousness, and the like are signs of progress but are not the "measure" itself. The power of the "measure," its honesty, stems from its refusal to accept an easing of human material existence or even an ecological truce in man's struggle with nature as the ultimate standard.

While the intent of the new science is clarified by seeing it as a "measure," it should not be overlooked that within the context of his larger project, the new science rests upon a contradictory foundation. Indeed, by virtue of Marcuse's preference for freedom over reconciliation the new science contains the seeds of its own abandonment. Marcuse's failure to pursue the new science further seems to stem in large measure from his pursuit of an even more radical project outlined in *Eros*: the replacement of reason with gratification. Because Marcuse's goal is not to experience pleasure from thinking well—i.e., the goal is not to make a life of reason pleasurable in the traditional sense but rather to experience pleasure *simpliciter*—the following strategy readily suggests itself: to let science and technology remain incompletely rational instruments, so that they can go about what they have shown themselves to be so good at—the material transformation of nature. Man could thus use the time won from labor to enjoy himself per se. The culmination of reason is its dissolution in gratification. Science and technology—in this Marcuse's fundamental strategy of freedom—serve reason by bringing about the necessary condition of its dissolution: freedom from the labor imposed by nature's scarcity.

In "The Philosophical Foundation of the Concept of Labor in Economics" (1933), Marcuse regrets the burden of labor and goes on to write of the analogous "labor" of human self-realization. Twenty years later in *Eros and Civilization* Marcuse separated labor and self-realization utterly, and has since gone on to make the abolition of labor the centerpiece of man's evolving triumph over nature.[46] With sufficient time for self-development, the world need no longer be an alien place for man. It is nature's demands upon man's time and energy that account for his experience of it as a place in which he must constantly deny his own needs in order to labor. It is this self-denial that in turn leads to that process critiqued by Horkheimer and Adorno, in which the domination of nature requires that man dominate his own nature.

The conclusion seems obvious, although it is never explicitly drawn by Marcuse. Clearly, though, the implicit logic of Marcuse's argument is as follows: if one removes the need for incessant labor, the process characterized by Horkheimer and Adorno might be circumvented at the start. There would be no need for "monistic" strategies of reconciliation with nature on the basis of a higher, more intuitive encounter with nature, for the repression of the erotic aspects of self that this encounter is designed to overcome (at least in Marcuse's scheme) would not occur in the first place. With this consideration, much of what is otherwise so puzzling about the new science falls into place. The new science is rhetoric designed to soften Marcuse's otherwise terribly harsh—especially in light of Horkheimer and Adorno's original goal of reconciliation with nature—goal of the complete subordination of nature to human purposes. The new science is, in a sense, an ideology. It grants the aura of reconciliation with nature to what is actually projected to be humanity's final victory over it.

The Aesthetic Dimension Reconsidered

The Aesthetic Dimension (1978) is sometimes interpreted as a disappointing retreat from practice by an aging and embittered radical. Marcuse's preoccupation with aesthetics in his last book has troubled even sympathetic critics.[47] Perhaps Marcuse was all along more interested in art than politics, one might conclude after reading this book. However, the thesis of this chapter, that Marcuse in effect chose absolute freedom from nature over reconciliation with it several decades prior to *The Aesthetic Dimension*, casts this work in a new light. Actually, the seeds of Marcuse's abandonment of the new science are apparent somewhat earlier, in *Counterrevolution and Revolt*. Although *Counterrevolution* contains some of Marcuse's most ambitious rhetoric regarding the new science, it also contains a reinterpretation of the role of sensuous perception, the epistemological foundation of the new science. Marcuse explicitly rejects his earlier assertion that the aesthetic image "cancels" the ugly and sad by transforming it into beautiful form. Marcuse now claims that the aesthetic image "eternalizes" these aspects of reality.[48]* This shift has significant implications. Earlier, sensuous perception bypassed the (dialectically revealed) history of a thing in order to get to its core. The core is the essence of a thing: its one true form. The core is likely to be a source of pleasure and beauty, because the history of a thing that sensuous perception bypasses is overwhelmingly a history of human unhappiness. That sensuous perception no longer "cancels" or "redeems" the ugly and sad implies that it no longer has the task of uncovering the one true form of things. Sensuous perception now operates entirely within its own dimension: the world of the eternal standards of beauty—the aesthetic dimension. The new science becomes art. That is, sensuous perception, once the ground of the new science, is now the basis of the aesthetic dimension and operates in the autonomous realm of artistic form. Sensuous perception no longer transforms nature; it memorializes, among other things, the desirable aspects of nature. This was, in the last analysis, the strategy of reconciliation with nature adopted by Horkheimer and Adorno, in which man is reconciled to his own memory of nature.[49]

Marcuse still writes of reconciliation in *The Aesthetic Dimension*; what has changed is its object. No longer at stake is man's reconcilia-

*Three years earlier, in *An Essay on Liberation*, Marcuse wrote of the redeeming quality of art:

> The phrase is indicative of the internal ambivalence of art: to indict that which is, and to "cancel" the indictment in the aesthetic form, redeeming the suffering, the crime. This "redeeming," reconciling power seems inherent in art, by virtue of its being art, by virtue of its form-giving power. (p. 43)

Three years later art no longer "cancels" or "redeems" suffering. Art eternalizes it.

tion with nature, but his reconciliaton with himself. Such reconciliation with oneself may be called catharsis.

> The aesthetic form, by virtue of which a work stands against established reality, is, at the same time, a form of affirmation through the reconciling catharsis. *This catharsis is an ontological rather than psychological event.* It is grounded in the specific qualities of the form itself, its nonrepressive order, its cognitive power, its image of suffering that has come to an end. But, the 'solution,' the reconciliation which the catharsis offers, also preserved the irreconcilable.[50]

It is in catharsis, in which man accepts that which cannot be changed, that he comes to terms with the ultimate duality of man and nature, subject and object. Man may make nature more tractable, he may bend it to his purposes through science and technology, but he may not—it appears—make it more truly human. That "the world was not made for the sake of the human being and it has not become more human" is the regrettable fact that man comes to terms with in the aesthetic catharsis.[51]

In claiming that the aesthetic catharsis is ontological, not psychological, Marcuse may be read as suggesting that in the world of art, in the world of our imagination and deepest longings, the ultimate reconciliation with nature remains possible. At the same time, the form of art—characterized by its existence outside of history, possessing its own independent reality—tells us that this reconciliation is not to be realized in this world. The artistic form of the promise cancels the promise but not the hope that gave birth to it. The cancellation of the promise is thus not trivially affirmative; it does not take the form of "But we knew it was impossible all along." The independence of the form of expression retains an element of negation. The promise: "es war doch so schön!"[52] The truth of the reconciliation with nature as the furthest achievement of human freedom and happiness remains, but it will not be realized in this world, at least not soon.

Whether Marcuse would have indeed abandoned the new science even as a rhetorical strategy must, of course, remain speculation. However, an essay written shortly before his death, "Protosocialism and Late Capitalism," provides grounds to think he might have.[53] The essay is a commentary on Rudolf Bahro's *The Alternative in Eastern Europe*.[54] In it Marcuse returns to some of his earliest themes. The "de-verticalization" of the division of labor could so reorganize necessary labor time that the realm of freedom might indeed, says Marcuse, appear within the realm of necessity.[55] In this vein Marcuse discusses the "economy of time" [*Ökonomie der Zeit*] in a manner strikingly reminiscent of his 1933 article on labor.[56] Yet, though Marcuse has an opportunity to do so—the door is opened to him, as it were, by such themes—he does not even hint in the direction of the new science. Instead, he analyzes alternative political strategies by which such goals might be attained. He embraces the activities of "catalyst groups," which include concerned scientists, the women's liberation movement, and *Bürgerinitiativen* (citizens' initiatives).[57] Furthermore, unlike his search for the "subject of the revolution" in several previous works, he no longer seeks simply the spontaneous fragmentary manifestation of the new sensibility. Rather, he analyzes the process by which it may be developed and made politically relevant on a "local and regional" basis.[58]

It is as though Marcuse finally comes to terms, in *The Aesthetic Dimension*, with some dualities that cannot be unified in this world but only in the world of art. Instead of driving him into pessimism, this seems to free him to attach his vision of freedom—which remains freedom from labor—to actual political possibilities, many of which are catalogued in Bahro's book, which is itself something of a list of utopian (but in principle quite possible) possibilities.

The Persistence of the Enlightenment Ideal of Science

Marcuse embraces much of the Enlightenment ideal of science, although he does so in a convoluted fashion under the influence of the critique of the dialectic of Enlightenment. Ironically, it is science as an instrument of the domination of nature that he ultimately finds most liberating. The Enlightenment ideal of science that he embraces is that of the utopian popularizers of science, such as Francis Bacon and Thomas More. To be sure, he does not share their optimism, so characteristic of the Enlightenment, in the potential of scientific reason to "migrate" throughout society and so make other social institutions more rational. Rather, it is Bacon's and More's embrace of science as the vehicle of man's liberation from the bondage of sickness, toil, and scarcity that he adopts.

Lord Macaulay's clever comment on Bacon might well be considered praise by Marcuse, though certainly not by Habermas: "The aim of the Platonic philosophy was to raise us far above vulgar wants. The aim of the Baconian philosophy was to supply our vulgar wants."[59] Marcuse would, of course, reinterpret "vulgar wants" as true needs, but he does consider Plato's rationalization of pleasure to be a detour from genuine gratification, a detour that may onc day be eliminated. For the first time in history, science may make it possible to provide man with enough. He would thus no longer need to construct elaborate rationalizations—such as idealist philosophies for the elite and religion for the masses—in the face of scarcity, toil, and unhappiness. No longer would pleasure in thinking be necessary. Humanity could finally experience pleasure per se.

CHAPTER 5

Habermas: Science and Survival

While Marcuse's analysis is often epigrammatic, Habermas' is stringently systematic. Yet, this does not always make Habermas' project easier to analyze. One reason this is so is because he has not one system, but several. Yet, through all the changes one theme has remained constant: the attempt to circumscribe the domain of science by means of the a priori determination of the limits of various types of knowledge vis-à-vis their objects. The tension in Habermas' project stems from this attempt, not entirely unlike Marcuse's, to absorb more and more elements of modern science while still retaining fundamental distinctions between types of knowledge. Indeed, it is this tension that has been the motor of Habermas' project for more than fifteen years, leading first to a more complex system necessary to save the hypothesis of cognitive interest theory and finally to what is really a new system: the reconstruction of cognitive competencies.

Habermas has sought to circumscribe science in several ways. Cognitive interest theory attempts to circumscribe science by establishing critical theory's epistemological independence of it. It does so by positing that language is the medium of human emancipation, while showing that because language constitutes the conditions of science language cannot be fully scientifically explained. In his more recent work Habermas has employed a different strategy. Critical theory itself becomes scientific. Habermas seeks to overcome the conventional character of critical theory—and the premature identification of reason and will represented by the emancipatory cognitive interest—by founding critical theory upon nomological universals revealed by a new "reconstructive science." Habermas uses science to found a critical theory that limits what science may explain. The next four chapters establish the continuity between these strategies. In *Theorie des kommunikativen Handelns,* Habermas' tone regarding science is somewhat different. The "new philosophy of science" understands itself so modestly, Habermas now believes, that it need no longer be epistemologically contained. Yet, this interpretation of science is not well integrated in Habermas' system. *Theorie* does, however, mark a significant departure in Habermas' work. In chapter 8, Habermas' remarks about science in *Theorie* will be considered. The more general implications of his revised view of science for issues concerning nature and the environment will be examined in chapter 9.

A Non-Scientistic Philosophy of Science

In 1971, Habermas argued in "Wozu noch Philosophie?" ("What is the Purpose of Philosophy?") that the task of contemporary philosophy is to criticize science in a way that does not destroy it. "If there should be a philosophy, in view of which the question "What is the purpose of philosophy?" no longer arose, it would, in view of our considerations, today necessarily be a non-scientistic philosophy of science."[1] A "non-scientistic philosophy of science" would keep science in its epistemological place without denying its achievements. Should philosophy not come to terms with modern science, philosophy risks being driven into irrelevance by the marvelous achievements of modern science. However, a philosophy that seeks to come to terms with science risks being absorbed by it. Such a philosophy would abandon its traditional commitment to a truth that stands beyond mere agreement with the empirical present. "The Classical Doctrine of Politics in Relation to Social Philosophy" is Habermas' most ambitious early attempt to walk this fine line.[2]

The promise of classical politics, says Habermas, was practical wisdom and prudence (phronēsis); i.e., right orientation in contingent circumstances.[3] The classical philosophers, by whom Habermas apparently means Plato and Aristotle, understood that politics could never be a science; rather, their political philosophy provided ethical guidelines for appropriate action in a changing and uncertain world. Through debate and discussion—which Habermas calls rhetoric—classical political theory taught men to act in a manner ethically becoming for human beings. However, says Habermas, modern political theory, which he sees as beginning with Machiavelli and Thomas More, promises more: it promises technically correct solutions. Whether Habermas is correct in his interpretation of the difference between ancient and modern political thought is not really at issue here, although one point jumps out to such a degree that it can hardly be ignored.

It is well known that Leo Strauss in effect argues precisely the opposite of Habermas. For Strauss the difference between ancient and modern political philosophy is that modern political philosophy promises *less*.[4] Habermas and Strauss are in rough agreement as to what modern political thought promises: security, order, stability. Strauss, though, sees this as a lowering of the sights of modern political thinkers, because he focuses upon an aspect of classical thought that Habermas surprisingly ignores. The classical thinkers, says Strauss, sought more than political prudence. They sought (even if they were not always sure they had found) objective moral and political standards. Habermas, on the contrary, states that "Aristotle emphasizes that politics, and practical philosophy in general, cannot be compared in its claim to knowledge with a rigorous science, with the apodictic episteme."[5] Habermas is not completely mistaken, of course (see Aristotle's *Nicomachean Ethics*, 1094b10–1095a1). However, Habermas' account is surely one-sided. The most famous piece of ancient political theory, Plato's *Republic*, is nothing if not an attempt to base politics upon episteme (519a–521c). This is the case whether or not one reads Plato's program literally, as the otherwise quite different readings of Strauss and Karl Popper testify.[6] Why would Habermas ignore this otherwise hard to ignore aspect of ancient political thought? Apparently because he wishes to show that the quest for certainty, with the consequent deemphasis on discussion and rhetoric, is an exclusively modern phenomenon associated with the rise of a protoscientific way of thought, exemplified by the works of Machiavelli, More, and Hobbes.

It is certainly the case, Habermas continues, that Machiavelli and

More were not social scientists. Machiavelli and More could hardly draw upon modern science, in which abstract theoretical knowledge is used to produce accurate predictions and hence technically usable knowledge. Modern science only arose with the method of Descartes and the mechanics of Newton.[7] Nevertheless, says Habermas, what makes the approach of Machiavelli and More scientific is their use of abstract models in the service of survival. Now, it would seem that More's *Utopia* is nothing as much as rhetoric. It is satirical exaggeration designed to publicize flaws in an existing society by comparing it to an ideal fictional one. Marcuse's new science may be seen in similar terms: the repressive uses to which science has been put are dramatized by its potential as a vehicle of absolute freedom. Were Habermas to view *Utopia* in this fashion, its continuity with the "Classical Doctrine" would be apparent. Instead, Habermas views More's book as a fiction with the attributes of a controlled empirical situation.[8] For Habermas, *Utopia* is a subjunctive counterfactual conditional hypothesis: "Should men be placed in such a situation, then they would act like so." Machiavelli, while not creating such an elaborate fiction, also abstracts from the ethical context of action, states Habermas. The acquisition and maintenance of power is the only concern.

> Just like the techniques for securing power in Machiavelli, so in More the organization of the social order is morally neutral. Both deal not with practical questions (i.e., ethical questions), but with technical ones. They construct models, that is, they investigate the fields, which they themselves have newly opened up under artificial conditions. *Here even before experimental methods were introduced in the natural sciences, methodical abstraction from the multiplicity of empirical conditions was initially tried out.* Surprisingly, even in this respect Machiavelli and More occupy the same level if one assigns the same heuristic significance to the former's 'realpolitical' demystification as to the latter's Utopian construction.[9]

In the politics of Machiavelli and More, suggests Habermas, ethics and values are detached from the protoscientific study of the fundamental threats to survival: the physical threat of the enemy or of hunger. It is as though modern *social* science arose roughly a century before modern science, with schemes that replace education in the terribly ephemeral attributes of prudence and wisdom with the promise of technically efficient solutions to the threats of starvation or political intrigue. In so doing, suggests Habermas, these schemes effectively reduce all values to the lowest common denominator of survival.

It is the purpose of "The Classical Doctrine of Politics" to lay out a background to the so-called "positivist dispute in German sociology," which Habermas entered in the same year as the publication of "The Classical Doctrine" with his essay "The Analytical Theory of Science and Dialectics."[10] The latter essay was in large measure a defense of Adorno's concept of the dialectical totality. Evidence of Habermas' intention in "The Classical Doctrine" is readily found. He refers to the "decisionistic" character of the modern philosophy of science twice.[11] "Decisionism," and the possibility of overcoming it, was perhaps the fundamental issue in the "positivist dispute." In addition, Habermas refers to Karl Popper's *The Logic of Scientific Discovery* as an example of scientific self-understanding that has completely abandoned the legacy of classical politics.[12] To be sure, there is nothing wrong with a partisan history of "scientism." The problem with this essay lies deeper: in its use of, in Habermas' term, a "stylized" history of political philosophy to enunciate permanent principles of scientific inquiry, such as Habermas' assertion that man's scientific interaction with nature is "in principle" [*im Prinzip*] "lonely and silent."[13] The character of science should rather be addressed it would seem by a philosophy—broadly interpreted to include the results of science—concerned with what is, rather than with a reconstruction of what several political philosophers have thought. This consideration extends beyond this particular essay to include much of Habermas' early cognitive interest theory, but not his theory of social evolution.

The "Classical Doctrine of Politics" argues that the attempt to scientifically understand the world is but a continuation of the attempt by Machiavelli and More to order the world technologically in order to meet the basic threats to survival. Machiavelli and More employ the experimental method in the service of survival one hundred years before this method became the centerpiece of the philosophy of modern natural science. In this vein, Habermas, quoting Hannah Arendt, argues that the guiding interest in the approach of Machiavelli and More already partakes of the essence of science: "'acting in the mode of producing.'"[14] Social philosophy and science both become virtually a mode of labor, or instrumental action. In fact, Arendt's view of science (which has striking similarities to Habermas'; indeed, Habermas has suggested that *The Human Condition* served as an inspiration for aspects of cognitive interest theory) perhaps more clearly than Habermas' recognizes the complex origins of science, especially its "inner affinity" to the contemplative tradition of Greek philosophy.[15]

Using Habermas' categories for a moment, one can say that Arendt distinguishes between two kinds of instrumental action: labor and fabrication, or work. She contrasts both to action and contemplation. Labor is man's "metabolism" with nature; it concerns the production of consumables, those things that must be continually grown or extracted, lest man perish. Fabrication, or work, produces the more durable human artifice: houses; factories; social, political, and economic institutions. Action is about human words and deeds. Arendt classifies science under the rubric fabrication, whereas Habermas—who makes no distinction between labor and fabrication—classifies science under the category of labor.[16] The virtue of Arendt's approach is that it seems to grasp the relationship between science and contemplation more clearly than Habermas' approach, which distinguishes solely between labor and language. Science and contemplation are related, states Arendt; they have an "inner affinity," because for the Greeks, contemplation—beholding something—was considered an inherent aspect of fabrication as well. The craftsman's work was guided by the idea, the model in his mind's eye, throughout the fabrication process, just as the philosopher was guided by the idea.[17] Arendt, like Strauss, in effect helps correct Habermas' tendency to sometimes read the history of philosophy and political thought too narrowly; i.e., strictly from the perspective of cognitive interest theory. Not only is the attempt to base politics upon certain knowledge considerably older than that of Machiavelli and More, but the relationship between this older tradition and science is, as Arendt's categories suggest, not quite as alien as Habermas implies.

In fact, Habermas is not really writing about science at all in "The Classical Doctrine." To be sure, he refers to science, but he is actually writing about social theory, the undesirable aspects of which, such as its ignorance of ethical prudence, he labels scientific. He is unconcerned, particularly in this essay, with science as a set of actual historical institutions or practices. Rather, he is concerned with an idea of science that posits that science—in order to wrest human existence from nature—must systematically ignore all but those aspects of reality that may be instrumentally manipulated. It is this interpretation of science as a pure instrument of survival that leads Habermas to so thoroughly blend social theory and natural science that he employs examples from the history of social theory to support claims about the very essence of science.

Decisionism.—Habermas' division of the world into two parts, those enterprises concerned with survival and those concerned with higher values, sharply influences his interpretation of the problem of "decisionism." "Decisionism" [*Dezisionismus*] is a multifaceted concept in critical theory.[18] It refers to philosophics that do not claim to be self-justifying, such as critical rationalism ("Popperianism"), as well as to those philosophies that glorify the act of deciding itself, no matter what its content, such as some strands of existentialism. Here decisionism refers to the "meaning gap" that science cannot fill, because science cannot speak to the significance and purpose of human existence. Science cannot tell us what the survival it serves means in human terms or even why survival is worthwhile. Yet, individuals require that such questions be answered. Into this meaning gap rushes a potpourri of mythology, arbitrarily justified values and beliefs that may or may not reflect progressive aspects of the culture, and ideology. In his later work Habermas focuses almost exclusively upon the steering of needs for meaning and significance by social systems based upon money and power. However, in his earlier work another threat is emphasized. One can hypothesize that this concern was stimulated by National Socialism. "Demons" come to fill the decisionistic gap. Demons are unleashed and uncontrolled needs and drives. If they remain repressed they can readily be utilized by an aggressive ideology that steps in to fill the decisionistic gap.[19] Science cannot exorcise these demons.

> Under the conditions of reproduction of an industrial society, individuals who only possessed technically utilizable knowledge, and who were no longer in a position to expect a rational enlightenment of themselves nor of the aims behind their actions, would lose their identity. Since the power of myth cannot be broken positivistically, their demythologized world would be full of demons.[20]

Only rhetoric, "which 'throughout deals with the audience to which it is addressed,'" can meet the needs for meaning and significance.[21]

By rhetoric Habermas means the offering of reasons for decisions coupled with the examination and discussion of these reasons. Rhetoric is the medium through which the classical doctrine of politics was practiced, says Habermas. Rhetoric treats its "object" as a partner in dialogue. In this respect its "logic" is the opposite of instrumental action. This is, of course, why Habermas so appreciates the prudential aspects of the classical doctrine of politics.

Habermas believes that science can make no contribution to exorcising the demons of decisionism. Earlier critical theorists, particularly Adorno, have seen science itself as a demon. Science remains too much a part of nature. It is too concerned with self-preservation and thus becomes a type of "wild self-assertion" [*verwilderte Selbstbehauptung*], in which man seeks mastery over everything in order to guarantee his survival.[22] Such an interpretation of the human "survival drive"—that it stems from an insecurity so profound that it cannot be quelled by anything less than absolute mastery—seems to owe more to Hobbes' speculations about human nature than it does to actual science.[23] What Habermas keeps of Adorno's interpretation is that science never really loses its roots in the basic survival drives, whereas rhetoric is concerned with higher values that are not merely an extension of self-preservation. Science itself is not a demon, in this view, but merely an instrument. However, science must be kept from reducing all values to the lowest common denominator of survival, lest the world be remythologized in the meaning gap left behind. It is this interpretation of science that is especially easy to read back into the schemes of Machiavelli, More, and Hobbes.

How sharply Habermas differs from Horkheimer, Adorno, and Marcuse on this whole issue is discussed in the introduction. Indeed, this is the issue that frames the debate between Marcuse and Habermas. Marcuse hopes that one day even an instrumental science will be absorbed into a comprehensive aesthetic-erotic orientation (even as much of his argument leads in another direction). Habermas believes that an instrumental orientation toward nature simply must be accepted as given (why he believes this is discussed in the next section). The trick is to prevent instrumental reason from becoming reason per se. Habermas does so by developing and defending a parallel concept of practical reason, which he claims is irreducible to instrumental reason. This strategy, as Joel Whitebook points out, accepts the disenchantment of the world as a necessary consequence of material progress and human liberation.[24] The natural world will always be an object, a thing from which we must wrest our existence, never a subject, never a partner.

The goal of Habermas' project can be expressed in one sentence: to prevent social relations from becoming like our relations with the natural world. Social relations are potentially transparent. Men and women have the potential to respond to each other as understanding subjects. The risk is that the power of science and technology will prove so attractive that this difference between humanity and nature will be forgotten. Habermas has stated that Horkheimer and Adorno's *Dialectic of Enlightenment* exerted a profound influence on him.[25] The influence obviously did not lead him to accept its analysis; rather, it seems to have led him to be enormously suspicious of the scientific explanation of human action: at any moment such explanation risks stripping men and women of their humanity. While this guiding insight has proven enormously fruitful in Habermas' work, it also contributes to the impasse his treatment of science runs into. Far too many different types of speculation that do not fit the categories of his system become instances of self-objectification. This issue is pursued in chapter 9 especially.

In the early 1960s, Habermas wrote as though Karl Popper's *The Logic of Scientific Discovery* (1935) were the culmination of the philosophy of science. Against Popper's fallibilism, as well as the logical positivism of the Vienna Circle, Habermas developed a philosophy of science loosely drawn from the pragmatism of Charles S. Peirce.[26] More recent developments in the philosophy of science, the so-called "new philosophy of science" of Thomas Kuhn, Paul Feyerabend, Imre Lakatos, and Mary Hesse, among others, have dated Habermas' early criticism of science. Indeed, Mary Hesse notes that there is now a "post-Kuhn and post-Feyerabend" set of debates within the philosophy of science, primarily associated with the work of Davidson, Kripke, Putnam, and others "who more or less indirectly owe their problem situation to the work of Quine." This, as Hesse suggests, places Habermas' early studies of science at least two generations behind current debates.[27] Habermas recognizes as much.[28] In *Theorie des kommunikativen Handelns*, as well as in his earlier "A Postscript to *Knowledge and Human Interests*," Habermas treats the new philosophy of science most sympathetically, though it will be shown that Habermas' attitude toward it is ultimately ambivalent.* For the most part, though, rather than taking up these new developments (even in *Theorie*, the new philosophy of science is a peripheral issue) Habermas has turned instead to criticism of especially egregious examples of technocratic social philosophy, such a Niklas Luhmann's systems theory.[29] However, in not always distinguishing his criticism of Luhmann, for example, from his criticism of scientism, Habermas perpetuates their misleading identification begun in "The Classical Doctrine of Politics."† (Habermas defines "scientism" as the exaggerated belief of science in itself; i.e., that it is the one true path to knowledge. However, the term has additional connotations for Habermas, such as that science is insufficiently reflective of its social origins and conditions of possibility; i.e., its transcendental preconditions. These additional senses of the term "scientism" will be discussed at various relevant points in the text.)

The Technical Cognitive Interest and Science

In "Labor and Interaction: Remarks on Hegel's Jena Philosophy of Mind," Habermas is concerned with much the same problem as in "The Classical Doctrine." However, in "Labor and Interaction" the outlines of Habermas' unique epistemological solution are apparent.[30] In Hegel's Jena lectures, according to Habermas, labor and social interaction are seen as irreducible to each other. The organization of society based upon norms has no logical connection with the causal processes that characterize man's relationship to nature.[31] To be sure, Hegel interpreted both labor and interaction in terms of his identity theory: in the end nature too finally becomes an expression of man's spirit.[32] However, Marx (who was apparently unfamiliar with the Jena lectures) was able to appropriate the con-

*Thomas McCarthy, in *The Critical Theory of Habermas*, notes that Habermas fails to update his criticism of the philosophy of science, pp. 60–61. However, one can consider Habermas' *Theorie des kommunikativen Handelns* a modest update. As early as 1973, in "A Postscript," Habermas stated that the new philosophy of science, which he calls the "confrontation of science theory with the history of science," has rendered the critique of scientism obsolete, p. 159.

†Actually, Habermas does draw a number of distinctions between science and systems theory. The problem is rather that until recently he rarely made these distinctions systematic and explicit. See chap. 8, pp. 119–23.

cepts of labor and interaction so as to help formulate his categories of the forces and relations of production.[33]

Habermas considers Marx's treatment of labor an improvement over Hegel's, because Marx, of course, abandons the Hegelian idealism. However, at the same time there is a fateful turn in Marx's analysis, according to Habermas. Marx devotes a great deal of attention to the actual empirical study of the relations of production; i.e., interaction. At the same time, Marx's materialistic philosophical system places labor at the very center of human existence. Though Marx analyzes social interaction separately from labor, his treatment of the problem of false-consciousness and the like takes place within a philosophical framework that suggests that ideology too can be seen simply as a reflection of how men labor. The reason this is the case, states Habermas, is that Marx (like Freud) misunderstood his theory to be a science.[34]

> Although he [Marx] established the science of man in the form of critique and not as natural science, he continually tended to classify it with the natural sciences. He considered unnecessary an epistemological justification of social theory. This shows that the idea of the self-constitution of mankind through labor sufficed to criticize Hegel but was inadequate to render comprehensible the real significance of the materialist appropriation of Hegel.[35]

The result—a combination of Marx's "scientistic self-misunderstanding," coupled with a materialistic world view that did not give as much philosophical credence to interaction as Marx actually gave it in practice—is the collapse of interaction into the category of labor. Reason becomes identified with the scientific study of society. Problems such as that of false consciousness come to be seen as technical problems, requiring merely the rearrangement of institutions and the like. This is, says Habermas, a tendency that Marx avoided quite well in practice but which is nonetheless fostered by his "scientistic self-misunderstanding." Later Marxists and most social scientists have not done as well.[36] Habermas' response to the incipient positivism in Marxism is in large measure epistemological: he seeks to demonstrate that knowledge of interaction is in fact a separate category of knowledge from knowledge of the material world.[37]

Habermas' "cognitive interest theory" is his alternative to the collapse of labor and interaction in Marxian epistemology. I focus upon the category of labor and its relationship to science. Habermas argues, as discussed in chapter 1, that "technology, if based at all on a project, can be traced back to a 'project' of the human species *as a whole*, and not to one that could be historically surpassed."[38] The term "project," it will be recalled, is Marcuse's. Furthermore, says Habermas, since the time of Galileo, science has been drawn into this species project: the *form* of the type of knowledge produced by both technology and science is "technically exploitable knowledge."[39] This claim has nothing to do with the intent of the scientist, which may well be to contemplate the truth of nature. It rather has to do with "the logical structure of admissible systems of propositions and the type of conditions for corroboration" of scientific theories.[40] As the passage from Stephen Toulmin quoted in chapter 1 (p. 13) suggests, such a claim—while not necessarily correct—need not be seen merely as an idiosyncrasy of "continental philosophy."

Key sources for Habermas' claim about this inner link between science and technology are Arnold Gehlen and, somewhat later, Jean Piaget.[41] However, Habermas' claim is actually based upon the category of knowledge he calls the "knowledge-constitutive interest in possible technical control" of nature. That Habermas' claim is so heavily dependent upon what amounts to his own reflection upon how men and women must see nature has not escaped criticism.[42] The cognitive interests, says Habermas, are "basic orientations rooted in specific fundamental conditions of the possible reproduction and self-constitution of the human species, namely *work* and *interaction*."[43] Cognitive interests stem from the things man must do to survive on this planet. However, while they stem from these material conditions, they do not stop there. Cognitive interests go on to constitute a transcendental (i.e., knowledge constitutive) horizon under which the world can be known. Habermas characterizes such a perspective as "quasi-transcendental."[44] This seems to mean that each interest sets the conditions under which the possible objects of experience can be known, but interests do not constitute the objects themselves. Habermas' is thus not simply an idealist philosophy. Quite the contrary. Habermas is claiming that the objects of experience, in conjunction with human nature (e.g., man must eat to survive), establish the framework within which these objects can be known. Indeed, some of the most trenchant criticism of cognitive interest theory, such as that of Thomas McCarthy, treats it as strictly an empirical hypothesis and asks what evidence, other than Habermas' own self-reflection, might support it.[45]

In "A Postscript to *Knowledge and Human Interests,*" Habermas summarizes the relevance of the technical cognitive interest for science: "The conjecture [is] that the use of categories like 'bodies in motion' . . . implies an *a priori* relation to action to the extent that 'observable bodies' are simultaneously 'instrumentally manipulable bodies.'"[46] Instrumental action is the type of action characteristic of the technical cognitive interest, and it is either successful in manipulating such "bodies in motion" or it is not. In *Knowledge and Human Interests,* Habermas suggests that the scientific experiment is simply a version of instrumental action. The successful experiment manages to manipulate the "bodies in motion" in accordance with expectations.[47] In "A Postscript," Habermas draws a distinction between instrumental action, on the one hand, and the discourse of scientists on the other. This distinction will be discussed shortly. However, it is interesting to note that in a 1982 essay on this topic Habermas reasserts that the only possible cognitive orientation toward nature is instrumental.[48] That is, scientists may seek to know the truth about nature, and in this sense may be said to pursue truth, not merely the control of things. However, this pursuit will necessarily remain within the confines of the technical cognitive interest. This means that the nature to which scientists refer as evidence for their theories will be accessible to them only under the horizon of the technical cognitive interest. There is no other alternative. Or rather, for Habermas there are two alternatives, neither of which is scientific. One is an aesthetic orientation toward nature, which while perhaps intensely gratifying does not produce knowledge. The other would be a regression "behind the levels of learning reached in the modern age into a re-enchanted world."[49]

Characterizing Habermas' Position on Science

Mary Hesse notes that Habermas rejects three positions commonly held by contemporary philosophers of science: (1) the *realistic interpretation* of science, under which its theories and laws represent actual entities; (2) that part of the *reductionist program* that claims that some day human action might be scientifically explicable in terms of natural processes; and (3) the *theoretical pluralism* of Paul Feyerabend, Richard Rorty, and others, which opens the door to attempts to reinterpret ordinary experience and language in terms of radically new constructed languages.[50] It should also be noted that Habermas rejects the *instrumental interpretation* of scientific theories (the distinction between an instrumental interpretation of scientific theory and the more fundamental instrumentalism of the technical cognitive interest is discussed on pp. 10–11). A definition and brief discussion of each of these four positions, as Habermas interprets them, is given in the excursus at the end of this chapter. How and why Habermas seeks to carve out a philosophy of science that rejects each of these positions is a guiding theme of the next chapter. The answer, expressed in the most general terms, is that Habermas seeks a certain stolid, though limited certainty from science. In this respect Habermas is not unlike Marcuse, who in his dominant strategy for human freedom embraces the blunt power of science, but not its claim to know the world.

In responding to each of these four positions, Habermas links the reductionist program with theoretical pluralism too closely. This is the most fundamental difficulty with his treatment of science, especially in his more recent works. By virtually equating reduction and pluralism, Habermas acts, in effect, to fix the categories of human knowledge for all time. He does so in the belief that to do otherwise is to open the door to rendering "the endangered humanity of the historical world compatible with the dehumanised reality of the objectivating sciences, which have been robbed of all human meaning."[51] It is, however, this broad equation of theoretical pluralism with dehumanization—the influence of the dialectic of Enlightenment is here especially apparent—that leads to a certain impasse in Habermas' project.

Excursus on Realism, Reductionism, Theoretical Pluralism, and Instrumentalism

To treat each of these topics fully would require a book on each. Here each is defined in a way that seems to fit Habermas' understanding of these concepts or that is at least relevant to Habermas' understanding of them. The purpose is to clarify Habermas' position. It is not to resolve difficult outstanding issues in the philosophy of science.

Realism.

> The *realist* view of laws is "realistic" in the technical sense which this term has in epistemology and ontology. In this view, a law is a universal and the relations of invariance which it asserts exist in nature independently of their being known, or of the conditions under which they come to be known. Knowledge of this universal is therefore a discovery of it.[52]

This view may apply to scientific theories as well.[53] In the realist view, the prevailing laws of science state what the discipline believes is its warranted belief about these laws. A version of realism asserts that any scientific law is a partial truth, relative to the evidence and the framework within which this evidence is significant. "Thus, the true propositions in which the laws of nature are asserted are objectively true, but our knowledge of them, at any one time, is relative. The fallibility of the laws of science lies, therefore, in this relativity."[54] This view shows especially clearly how, contrary to Habermas' interpretation (as discussed in chapter 6), scientists can reject one natural law for another without being seen to have put themselves in the untenable position of claiming that the laws of nature have themselves changed.

As is shown in chapter 8, Habermas can be seen as in effect asserting a realistic view of what he calls "reconstructive science." Reconstructions describe actual "laws of nature," in this case having to do with how humans cognitively and linguistically apprehend reality. In discussing reconstructions, Habermas agrees that they can be mistaken, like any other empirical hypothesis.[55] This would seem to support the claim that reconstructions should be interpreted as an expression of scientific realism. However, Habermas draws some further distinctions (discussed in chapter 8) that have the effect of clouding this issue. Nevertheless, when all is said and done his position regarding reconstructions seems to be in fact identical with a realistic view of scientific laws: Where "L" stands for the law of science, and L for the law of nature, realism states that "L" is true if and only if L.[56]

Reductionism.—Reduction "is the explanation of a theory or a set of experimental laws established in one area of inquiry, by a theory usually though not invariably formulated for some other domain."[57] Two somewhat different types of reduction can be distinguished. Habermas' objection is to the second type. In the first type of reduction, a theory originally formulated to explain quite a restricted class of phenomena may be extended to explain those phenomena even when they are manifested by a more inclusive class of things. A familiar example is the theory of mechanics, originally developed to explain the motions of bodies whose dimensions are negligibly small when compared to the distance between the bodies (point-masses). Later this theory was extended to rigid bodies. However, in this type of reduction substantially the same concepts are employed in formulating the laws in both domains.[58] The second type of reduction may occur between what seem to be entirely different classes of phenomena. The difficulty with this type of reduction is that the primary science (that to which something else is reduced) may seem to wipe out familiar distinctions and experiences, showing them to be spurious, epiphenomenal, and the like.[59]

Habermas is concerned that one day science may seek to explain language in scientific terms that seem to have nothing to do with the experience of using language, such as the terms of neurophysiology. This is what Habermas' charge of "self-objectification" against the "objectivating sciences" is about. Against this possibility Habermas may be seen as technically asserting the doctrine of *emergence*. This doctrine refers to characteristics of behavior (generally of groups, but the term is applicable to any phenomenon, including language) that are not predictable by laws referring to the elements of the group, including composition laws, which normally predict what occurs when individual elements of the "secondary" phenomenon interact. Actually, Habermas' position is more general and less precise. As in the debate over instrumentalism, Habermas is arguing at a somewhat different level: that even if one could scientifically predict linguistic events, this would not capture the meaning of language, especially the fact that by its very nature as language it is oriented toward mutual understanding, recognition, and truth. Russell Keat's discussion of reduction in psychoanalysis, in *The Politics of Social Theory: Habermas, Freud, and the Critique of Positivism*, is a helpful treatment of this entire issue as it pertains to Habermas.[60]

Habermas' position against reduction has in one respect received a great deal of support from the recent philosophy of science. For example, Paul Feyerabend has shown that an adequate "formal and 'objective'" account linking the phenomenon of a so-called primary theory with that of a so-called secondary theory, so as to demonstrate that meaning invariance has been preserved, has not yet been given.[61] As Feyerabend suggests, however, his should not be seen as an "in principle" argument against the very possibility of reduction. His is rather a challenge aimed at several cases of what are widely regarded to be successful reductions.[62] The basis of Feyerabend's argument is his position on *incommensurability*, a concept that has frequently been interpreted in an exaggerated fashion, as Harold I. Brown points out in his recent article on the topic.[63] Neither Feyerabend nor Kuhn believes that incommensurability means that theories cannot be compared.[64] What incommensurability does mean, at least for Kuhn, is that there exists "no common language within which two theories referring to different entities could be fully expressed and which could therefore be used in a point-by-point comparison between them."[65] It is the absence of such a common language that prevents the determination of whether "meaning invariance" has been preserved, and hence whether a particular reduction is valid. Feyerabend is even more tentative in his assertion of incommensurability, insofar as he treats it as a historical occurrence rather than a logical necessity.

The moral to be drawn is, I believe, that a certain skepticism toward purported type-two reductions, especially, is warranted. Such skepticism would presume unless shown otherwise (via the demonstration that meaning invariance has been preserved) that the purported reduction is actually a case of two different explanations of partially overlapping phenomena. Richard Rorty takes a similar position. While it might one day be possible to invent a new scientific language which would explain aspects of language, what one would be faced with—should this occur—would be two quite different accounts of language, a natural and an artificial account, but not a primary and secondary one.[66]

Theoretical Pluralism.—Theoretical pluralism is in many respects an outgrowth of the above considerations about reductionism, especially as developed in the work of Paul Feyerabend. If the facts rarely if ever falsify a single theory (as is widely recognized), and if so-called crucial tests between theories are plagued by the meaning variance problem (i.e., it is extraordinarily difficult to determine if two ostensibly competing theories are actually about precisely the same phenomenon), then "counterinduction" is the best strategy for the growth of knowledge. Counterinduction, according to Feyerabend, seeks to develop new theories that do *not* fit the facts, at least as these facts are described by accepted and highly confirmed theories.[67] The immediate goal of such a "counterfactual" procedure is the creation of a theory powerful enough to constitute new facts, facts that might begin to challenge the facts produced by an older, more firmly entrenched theory. For if theory *y* cannot falsify theory *x*, it may nonetheless make life uncomfortable for proponents of theory *x*, perhaps by predicting all sorts of new and interesting things. Knowledge, in this view, is characterized by the *proliferation* of theories and explanations. It is neither a series of consistent theories that converge on a single viewpoint nor is it a gradual approach to truth. Instead, knowledge is "an ever increasing *ocean of mutually incompatible (and perhaps even incommensurable) alternatives*, each single theory, each fairy tale, each myth that is part of the collection forcing the others into greater articulation and all of them contributing, via this process of competition, to the development of consciousness."[68]

This is the strategy of theoretical pluralism as stated by its most extreme proponent. There are other versions, all of which seem to stem from the "Duhem-Quine thesis," which states that there can be no definitive crucial tests between theories because "any theory can be permanently saved from 'refutation' by some suitable adjustment in the background knowledge in which it is embedded. As Quine put it: 'Any statement can be held true come what may, if we make drastic enough adjustments somewhere else in the system. . . . Conversely, by the same token, no statement is immune to revision.'"[69] It is in this context that Mary Hesse notes that "the post-Kuhn, and post-Feyerabend debates on truth and meaning, instrumentalism, realism and relativism . . . more or less indirectly owe their problem situation to the work of Quine."[70] It is ironic that Habermas so strongly opposes a view (theoretical pluralism) that was developed by its most aggressive proponent, Feyerabend, as an outgrowth of an argument about reduction, an issue over which he and Habermas are in basic agreement.

It is probably not fruitful to address the issue of whether theoretical pluralism is a form of relativism. Relativism has become such a loaded term that its usefulness in objectively characterizing a particular position seems minimal.[71]

Instrumentalism.

The central claim of the instrumentalist view is that a theory is neither a summary description nor a generalized statement of

> relations between observable data. On the contrary, a theory is held to be a rule or a principle for analyzing and symbolically representing certain materials of gross experience, and at the same time an instrument in a technique for inferring observation statements from other such statements. . . . More generally, a theory functions as a "leading principle" or "inference ticket" *in accordance with which* conclusions about observable facts may be drawn from given factual premises, not as a premise *from which* such conclusions are obtained.[72]

The difference between how instrumentalism is generally interpreted in the philosophy of science and what Habermas means when he states that the technical cognitive interest exhibits an instrumental orientation toward nature is addressed on pp. 10–11, and extensively throughout chapter 6. Against Habermas' position on instrumentalism, Hesse notes that while his objections based upon those arguments raised by Popper and others are sound, Habermas' argument "does not exclude the possibility of instrumental interpretation of higher-level theoretical postulates that remain underdetermined by the empirical." These postulates—even in the case of reconstructive science—may not be immune to cultural influences, and in this sense at least may be relative.[73]

CHAPTER 6

Habermas' Early Studies of Science and the Emergence of Language

HABERMAS' VIEW of science is best unpacked by beginning with his early studies of it. In these studies Habermas "solves" the induction problem. With this solution he explains what he sees as the most significant attribute of science: its cumulative and progressive character. Habermas asks:

> What are the properties of the transcendental conditions of a process of inquiry in whose framework reality is objectified such that we apprehend the general in the singular—that is, that we can infer the validity of universal propositions from a finite number of singular cases?[1]

The answer is the process of inquiry itself, which acts to structure reality in a definite constellation of universal and singular events.[2] Induction is a system of expectations built into the behavioral system of purposive rational action. We discover general principles in nature by acting as though nature itself behaves in a habitual manner. Scientific hypotheses stabilize our expectations, but only if we presume that nature itself seeks to stabilize its behavior by means of these same principles.

> Only if we attribute something like instrumental action to nature itself can we abductively discover new hypotheses, deductively derive conditional predictions from them, and confirm them through continued induction. We must act as though observable events were creations of a subject. . . . This subject would be nature, which had habitualized all 'laws of nature' as the rules of its behavior. Only if man in his instrumental action constitutes his natural environment from this point of view and projects himself as the opponent of an instrumentally acting nature can he hope for success with his method. . . . "This is our natural and anthropomorphic metaphysics."[3]

Actually, suggests Habermas, we create a feedback-loop between our actions and nature's response, but we must pretend, in order that new hypotheses can be discovered, that nature is responding to our actions. We must pretend that nature has "habitualized all 'laws of nature' as the rules of its behavior."[4]

Why must we make this assumption? Because the systematic discovery of new hypotheses, and thus the progress of science, cannot be adequately explained by the "randomness of a hypothesis-creating imagination" but only by the fact that falsifications "compel the abductive generation of new hypotheses and thus take on the form of *determinate negation*."[5]* Apparently what Habermas has in mind is that we must at least pretend that nature is a type of mirror-image partner in this dialectical process in order to explain scientific progress in an orderly fashion. I say "pretend" because Habermas' actual philosophic position seems to be that because nature cannot communicate, and thus cannot really take part in the

*In a footnote to "The Self-Reflection of the Natural Sciences," Habermas states that

> the *systematic* discovery of new lawlike hypotheses, however, freed from dependence on chance inspiration, is conceivable only if the unexpected result necessarily leads to a *determinate* negation, that is to a modification of the refuted lawlike hypothesis within the semantic domain it has already marked out (p. 334, n. 4; Habermas' emphasis).

It is difficult to tell if Habermas is just using the language of dialectics here, or if he is suggesting the process is itself dialectical. The original, *Erkenntnis und Interesse*, p. 147, n. 72, sheds no light on this question. Habermas' other work on this issue would lead one to believe he is just using the language of dialectics.

dialectic (Habermas understands dialectics overwhelmingly in terms of the dialectic of mutual recognition), nature remains essentially unknowable to us.[6] All we can know is the results of our own actions.

This perspective also "solves" the basis sentence problem. The basis problem, as formulated by Karl Popper, is that observational statements employed in the falsification of hypotheses are ultimately conventions. Their truth cannot be guaranteed by the fact of their simplicity or immediacy, as some earlier empiricists believed. Rather, scientists come to an agreement over what constitutes a basis statement in a process that resembles a trial: facts (evidence) are introduced in support of basis statements but only in the context of definite procedures. The procedures in this case are an expression of what we know about reality from our theories and everyday experience. This view has two interesting consequences. First, theories have an opportunity to determine what facts shall count against them. Second, our everyday experience, which is steeped in "value Platonism" (facts are seen as having values attached to them), enters into the evaluation of scientific theories. Science, it seems, really does rest upon a foundation of conventions. Yet, says Habermas, how are we to explain that contrary to the concerns of the philosophers of science we are usually in no doubt about basis statements? "The so called basis-problem simply does not appear if we regard the research process as part of a comprehensive process of socially institutionalized actions, through which social groups sustain their naturally precarious life."[7] The basis statement draws its validity from group convictions that have sustained themselves in action; i.e., that have worked. It is in this context that Habermas asserts that experiments "imitate" [*imitieren*] the feedback system that is built into the process of social labor.[8] Here the connection between the technical cognitive interest and science is close indeed. Habermas concludes, "All the assumptions, then, are empirically true which can guide feedback-regulated action without having been previously rendered problematic through errors experimentally striven for."[9] This is not an instrumental view of scientific theory but rather an instrumental interpretation of the scientific enterprise as a whole.[10] The methodological procedures of the community of scientists, who here represent the species that is interested in controlling nature, act to create a world that is stable enough to correspond to our expectations.[11]

It is tempting to see basis sentences as the link between science and valuation. Basis sentences are the media by which values are passed into science. Lorenz Krüger explicitly makes this claim in reference to Habermas' project.[12] In fact, this may be the most sensible way to interpret Habermas' otherwise puzzling claim that moral feelings, privations, frustrations, and individual crises can become instances of proof used to falisfy scientific hypotheses.[13] Yet this is not a fundamental aspect of Habermas' view of science. As Karl-Otto Apel points out, the value-constituted character of scientific facts seems to be canceled (Apel's term is *vorweggenommen*, or anticipated) by the instrumental character of science. Every fact science encounters is seen so overwhelmingly in light of its potential for manipulation that its other aspects are swamped.[14] Instead, the basis sentence problem is used by Habermas to support the claim that science always proceeds within the framework of the technical cognitive interest. The argument runs like this.[15] The Popperian philosophy of science (the philosophy of science, other than that of Peirce, that Habermas deals with almost exclusively during the period of his cognitive interest theory) recognizes the value-laden character of scientific inquiry. However, it attempts to push up all the values science is concerned with to the level of the values and ideals of scientific inquiry itself. For example, the philosophy of science recognizes the conventional character of basis sentences, but by considering this problem a philosophical one—something which threatens the objectivity of knowledge—it tends to make it abstract and unreal. Popper's metaphor of a scientific court of law deciding upon which facts will be accepted captures this abstractness. In fact, Habermas might well have said, a better metaphor would be the factory. The conventional character of basis statements reminds us that scientific facts are constituted not just by scientists and theories, but by the very real need of men and women to master their environment. The primacy of this need overwhelms any abstract interest of science in objectivity and truth. Much of the philosophy of science is simply a decoration of this circumstance.

The "traditional" philosophy of science falls out in this view, insofar as it is concerned with issues of objectivity and truth. It is replaced by the reconstruction of rules that define "the course of the objectification of reality necessary within the transcendental framework of processes of inquiry."[16] This reconstruction of the rules under which reality is objectified is characterized by a mixture of a certain dogmatism and skepticism. On the one hand, we can be absolutely certain of these rules because we grasp them by reflecting on what we have ourselves made, like a text from our own past.[17]

On the other hand, this is all we know, all we can know, and all we need to know. Nature itself, not the nature that we pretend acts out our hypotheses, remains forever closed to us, because it is not like some text we have written and forgotten, but simply the other with whom no reconciliation is possible. Habermas believes we may only truly know what we ourselves have made. "What raises us out of nature is the only thing whose nature we can know: *language*."[18] Hans Albert argues that Habermas, inspired by Heidegger, treats the world as a type of text that may truly be known only through hermeneutics.[19] However, this criticism is not quite correct. For Habermas the activities of scientists are indeed a type of text, but the world itself remains a blank sheet of paper.

The instrumental cognitive interest has replaced objectivity, "which makes a dogma of the prescientific interpretation of knowledge as a copy of reality."[20] It replaces the ontology of the factual, by which Habermas means the "illusion" that facts exist independently of our knowledge of them.[21] Yet, if we think of objectivity as signifying the commonsense notion that nature resists some intrusions but not others, Habermas' version does not seem to add much, with one significant exception. The resistant nature in question is ultimately reduced to human nature, in particular the conditions of human survival. In one sense, the "interest in instrumental action" is here just another name for objectivity. Objectivity has become the objective compulsion of interests.[22] This view is supported by Reinhart Maurer's observation that we appear to have no choice but to follow interests. Reflection upon them seems merely to be an expression of interests rather than to be constitutive of them.[23] Even discourse, examined shortly, does not change this. Maurer states that "self-reflection is an expression, or self-presentation, of interests, rather than knowledge of them, and discourse among all is their [interests'] reciprocal self-mediation."[24] While Maurer's assessment is on the right track it is too harsh. It fails to make several important distinctions, which will be raised in chapter 9.

The substitution of interests for objectivity is, however, not without consequence. It leads to a combination of dogmatism, skepticism, and subjectivism regarding our knowledge of reality.

> The universal matter of fact of the "hardness" of an object called diamond has existence (if and as long as there are diamonds) independently of whether any person actually makes the attempt to rub any diamond with the aid of a sharp object.

> On the other hand, it is not meaningful to attribute the predicate "hardness" to an object called diamond if the statement cannot be made at least implicitly with regard to the system of reference of possible instrumental action. We do reckon with the existence of a reality that is independent of men who can act instrumentally and arrive at a consensus about statements. But what the prediction of properties catches "of" this reality is a matter of fact that is *constituted only* in the perspective of possible technical control. It is in this sense that I understand the resolution of the paradox suggested by Peirce in his "Issues of Pragmaticism."[25]

Habermas is drawing a distinction between the validity of science, which can only be instrumentally understood, and the objectivity of the natural world as we ordinarily understand it.[26] The validity of science, suggests Habermas, cannot be separated from the way in which it constitutes the world as an object of instrumental action. In other words, science cannot apprehend the world in any terms but the terms of instrumental action. Consequently, the validity of science cannot be judged by any criteria except its success in generating feedback from nature that is in accordance with expectations. However, this does not mean that the natural world itself, the world of everyday life as we ordinarily experience it as simply "there" or "given," obtains its existence or essential characteristics from the constitutive activities of science. This is presumably the point of Habermas' designation of interests as having "quasi-transcendental" status.[27] Interests constitute all we may know about the world, but they do not constitute the world itself. Yet, it is precisely because interests do not constitute the world itself that the world must remain unknowable. This last point is the obverse of Habermas' claim, quoted above, that we can only truly know what we have made: language.

"A Postscript to *Knowledge and Human Interests*"

A recent (1982) exchange between Habermas and Mary Hesse helps to explain Habermas' concept of science more fully, particularly as he revised it in "A Postscript to *Knowledge and Human Interests*." In "A Postscript," Habermas develops the concept of discourse that he uses to distinguish more fully between instrumental action and science (see the note on page 100 for a brief summary of the formal characteristics of discourse). Scientists are distinct from laborers, Habermas now states, because scientists seek truth, not merely the most efficient way to manipulate "bodies in motion." Scientific discourse is isolated from reality. The constraints generated by reality arc rendered problematic in discourse. For example, scientists may seek to falsify an explanation that predicts perfectly well. It is in this context that Habermas modifies his earlier assertion that the scientific experiment "imitates" the feedback from instrumental action. Rather, instrumental action and scientific experiment are "structurally analogous actions pertaining to the realm of life-praxis and operations pertaining to discourses."[28]

Hesse focuses upon what she regards as a "somewhat opaque paragraph" in "A Postscript," which concludes that "it is therefore more plausible to assume that the objectivity of experience guarantees not the *truth* of a corresponding statement, but the *identity* of experience in the various statements interpreting the experience.[29] Hesse unpacks Habermas' claim by parsing it into five assertions.[30] (1) Since even the most basic observation statements are expressed in terms of one theory or another, and since these theories change over time, what it would mean to say that these statements correspond to reality is entirely unclear (the "basis sentence" problem). (2) Theories and their languages thus cannot be said to describe the world; rather, they interpret it more or less adequately. (3) Adequacy may be measured by experiment, but given the fundamental role of interpretation, the experiment can never be separated from argument about its relevance. (It is this elevation of the role of argument that characterizes Habermas' discussion of science in "A Postscript".) (4) Indeed, if adequacy were measured by empirical verification alone, we could never perform what Popper calls "crucial tests" between theories, for each theory would define the relevant phenomena in different ways, according to its unique assumptions (the "meaning variance" problem). (5) Therefore, a common reservoir of shared interpretations is needed, lest there be no common experience against which to test theories. This common experience is provided by the "technical cognitive interest," which guarantees that each member of the species confronts nature from within the same framework.

In his response, Habermas agrees with Hesse's interpretation.[31]

> Prior to all theory, there is constituted in the behavioural system of instrumental action a basis in the life-world for possi-

> bilities of reaching understanding; this basis explains why we can *refer to the same phenomena* through the lenses of *different theories*, even if these phenomena, when transformed into data, have to be described in connection with those very theories.[32]

Like Marcuse, Habermas seeks not merely to establish but also to fix the roots of science in the lifeworld. Unlike Marcuse, he does not do so in order to hold out the possibility of a new science based upon a new experience of the lifeworld. Rather, Habermas does so in order to limit the proper domain of scientific explanation by attaching it to a particular perspective on the lifeworld characterized by the technical cognitive interest. While the development of Habermas' treatment of science is characterized by a loosening of the link between science and the technical cognitive interest, the link is never abandoned, as Habermas' response to Hesse indicates. One difficulty with Habermas' view is that it seems to foreclose the possibility that utterly new categories of experience might emerge as the result of new scientific discoveries. If the lifeworld constrains our experience of nature as severely as Habermas suggests, then new scientific discoveries will at most reinterpret the lifeworld but never radically revise it. However, since the lifeworld is presumably not merely given but rather is a humanly constructed interpretation of experience, Habermas' position seems unjustifiably inflexible. As the preceding quote from Maurer suggests (p. 92), this rigidity in Habermas' position seems to stem from the transcendental character of cognitive interest theory. However, chapters 8 and 9 show that this charge of a certain rigidity in Habermas' position also applies to his more recent "reconstructive science," which is no longer based upon cognitive interest theory.

The elements of Habermas' interpretation of ordinary science are apparent. His interpretation of science is incompatible with scientific realism (point 1). Indeed, Habermas argues that realism makes no sense under his scheme. How could scientific laws and theories describe reality if the history of science is the replacement of one theory by another that claims to explain the same phenomenon? This would put us in the absurd position, says Habermas, of asserting that the laws of nature have themselves changed.[33] In defense of realism, one might argue that what was once regarded as warranted belief about a law of nature comes, in the situation Habermas describes, to be seen as unwarranted belief, to be replaced by a new scientific law that is currently regarded as warranted. A statement of a law of nature can be false without rendering the idea of independently existing and potentially discoverable laws of nature nonsensical. In any case, while Habermas rejects the realistic view of scientific theories, a more fundamental realism can be seen as characterizing his scheme. All scientists, indeed all men and women, apprehend nature under similar constraints, those of the technical cognitive interest. The constraints generated by this interest are broad and flexible. Nevertheless, the outer limits of these constraints can be described, and known, in a realistic manner. For the constraints of the technical cognitive interest are the conditions of human survival.

Habermas also denies the instrumental interpretation of scientific laws and theories. This is so even though Habermas believes that from a transcendental (knowledge constitutive) perspective scientific knowledge exemplifies an instrumental—i.e., "monologic"—approach to reality. Habermas rejects the instrumental interpretation of scientific theories on the basis of two considerations. First, scientists do not directly confront nature (point 3). They do so from the perspective of a detached discourse, in which other considerations are relevant. For instance, scientists may seek to falsify a theory that predicts perfectly well (i.e., is a perfectly good instrument) because they are interested in other values, such as truth or elegance. At this level, Habermas accepts the arguments against instrumentalism raised by Popper and others.[34] There is, however, a more fundamental reason that theories are not seen as mere instruments by Habermas. That some theories work, whereas others (e.g., perpetual motion theory) do not, reflects for Habermas that there is an objective reality with which our theories are in contact. Indeed, Habermas' distinction between the objectivity of science and its validity serves the purpose of allowing him to uphold this claim, while denying that science tells us about nature in itself. For Habermas, scientific theories may not describe nature in itself, but they are not mere "inference tickets" either. This is demonstrated by Habermas' response to Niklas Luhmann, a systems theorist who defends what is basically an instrumental view of science. Habermas responds that nomological statements (i.e., law-like generalizations) "have a function for the context of instrumental action becuse they are true; they are not true because they could have such a function."[35]

Given the centrality of the technical cognitive interest to Habermas' interpretation of science, it is clear why he would reject realism. But why would he argue so strenuously against instrumentalism? After all, the instrumental interpretation of science would seem to be especially compatible with Habermas' goal, which is to show that science can never escape the framework of the technical cognitive interest. A little speculation is perhaps in order. A strictly instrumental interpretation of scientific theory, Habermas seems to believe, would no longer see itself bound by the technical cognitive interest either but would seek to make predictions about language and other human phenomena. If scientific theories are mere instruments, then there is no obvious limit to their proper use. Claiming to be no more than an instrument, science might also approach human action in a strictly instrumental fashion. The ultimate result of such an approach to human action, Habermas suggests, would be "self-objectivation." How Habermas reaches this conclusion is discussed below.

The philosophical foundation of the possibility of "self-objectivation," according to Habermas, is the assertion by Paul Feyerabend, Wilfrid Sellers, Richard Rorty, and J. J. C. Smart, among others, that the reflexive language of everyday life could be replaced by theoretical languages. In particular, some scientists believe that scientific argument itself will one day itself be empirically explicable, and this would demonstrate that it was all along a category mistake to look for the transcendental foundations of science elsewhere.[36] (It should not be overlooked that none of the individuals Habermas cites in this regard are scientists; most, though certainly not Feyerabend, are analytic philosophers.)*

> This would make possible the progressive self-objectivation of speaking and acting subjects so that one day even the objectivistic self-interpretation amongst members of the scien-

*That Habermas is referring here primarily to the claims of some analytic philosophers may help to explain what is otherwise a puzzling aspect of "A Postscript": he praises the "new philosophy of science," but rejects a position closely associated with it—theoretical pluralism. This may be explained by the fact that in "A Postscript," the new philosophy of science is seen by Habermas as expressing the "confrontation of science theory with the history of science" (p. 159). Thomas Kuhn, whom Habermas exempts from the charge of "self-objectivation," would thus epitomize the new philosophy of science, whereas Rorty et al. would be seen as representing analytic philosophy. However, as Mary Hesse suggests, this is a tenuous distinction ("Science and Objectivity," p. 100). *Theorie des kommunikativen Handelns*, it will be shown in chapter 8, conceives of the new philosophy of science in somewhat different terms: as the coming together of the philosophy of science and hermeneutics.

tific community would become a possibility (displacement hypothesis).[37]

Habermas' is a puzzling claim. He seems to equate the scientific *explanation* of human action with the *replacement* of the categories of everyday life by scientific categories. The program of Rorty hardly fits this interpretation. As Richard Bernstein points out in *Praxis and Action,* Rorty simply asserts that it is possible that in some areas of life a constructed language might be more useful than a natural one.[38]

Nor does Feyerabend's program fit Habermas' interpretation. Though the terms "self-objectivation" and "displacement hypothesis" suggest that perhaps he does, Feyerabend does not pursue the classic reductionist program, as the conclusion of the last chapter points out. Quite the contrary, in fact, is the case. Feyerabend came to prominence with his assertion in "Against Method" (1970) that due to the difficulty of establishing meaning invariance across theories, the possibility of ever fully explaining one theory in terms of another ostensibly more fundamental theory is in doubt.[39] The issue is similar to that of the problem of so-called crucial tests between theories, addressed earlier in this chapter. Because facts are constituted by theories, theories have an opportunity to determine what facts shall count against them. The facts of theory *x* are likely to neither confirm nor deny theory *y* but simply be incommensurable with it. Given that this is the case, the approach most conducive to knowledge, Feyerabend concludes, is to let a thousand theories bloom. Knowledge, in this view, is characterized by the proliferation of theories and explanations. Knowledge is "an ever increasing ocean of mutually incompatible (and perhaps even incommensurable) alternatives . . . all of them contributing, via the process of competition, to the development of consciousness."[40] This is the position of theoretical pluralism, discussed previously. What Richard Rorty in *Philosophy and the Mirror of Nature* calls "edifying philosophy," which seeks to keep the conversation going rather than to find objective truth, is similar in many respects and is discussed in chapter 9.

Against theoretical pluralism in science, Habermas sets the technical cognitive interest as the source of humanity's common experience of nature. If such a common experience is available to be called upon in discourse, the problem of incommensurability can be contained (point 5, Hesse's interpretation). Rorty and Feyerabend in effect challenge this strategy. If entirely new languages referring to casually invented entities can be freely constructed, on what grounds can it be claimed that the technical cognitive interest circumscribes the proper domain of science? Habermas' assertion that natural science should generally restrict itself to the explanation of bodies in motion and the like would have to be seen as an ethical belief or preference rather than as an epistemological argument. He could no longer base his argument upon certain "quasi-transcendental" limits to the scientific experience of nature, as defined by cognitive interest theory. Theoretical pluralism challenges the foundation of Habermas' attempt to limit the proper domain of science: the claim that the technical cognitive interest is the common denominator of man's scientific encounter with nature.

In light of these considerations, why Habermas would be ambivalent toward the new philosophy of science is also apparent.[41] While he admires its epistemological modesty, it is not modest in quite the way Habermas would have it be. The new philosophy of science, at least as represented by Kuhn, Feyerabend, and Rorty, is modest insofar as it believes that reality is sufficiently manifold and ephemeral so that no single theory can definitively account for any aspect of it. However, the modesty that Habermas seeks is somewhat different: that science be modest about *what* it may properly explain; i.e., its domain of validity. It is the failure of the new philosophy of science to be modest in this regard that helps to explain that while Habermas praises the new philosophy of science he does not adopt its pluralistic account of science. Habermas rather retains his original account, in which science is a much more stolid and unimaginative enterprise, concerned with generating useful feedback from nature.[42]

The Emergence of Language

If the "objectivating sciences"* were to explain language, what might they discover? This is a relevant question. Habermas' is not

*Mary Hesse seeks to clarify the meaning of Habermas' term "objectivating sciences." She notes that

> to "objectivate" experience, then (at least as this term is used by Habermas' translator in the "Postscript"), is to take experience as the

only an abstract philosophical argument against the scientific explanation of language; his is also a practical defense of his own system. Consider the following example. The ideal speech situation (which Habermas employs in order to overcome decisionism) assumes that under special conditions—the general symmetry conditions of discourse*—individuals will be able to suspend their own private interests in favor of the general interest. That Habermas employs the ideal speech situation as a "counter-factual hypothesis," meaning that it need not be actually realized in order to be a valid standard, is irrelevant. Discourse must be *in principle* realizable in order to be a useful standard, even if Habermas never expects it to be actually realized. Were "ideal speech" unrealizable in principle, there would be no reason to take it seriously. Like Thomas More's *Utopia*, it would be no more than a fascinating literary construct. In fact, Habermas intends that "ideal speech" be in principle realizable. This is the point of his theories of communicative competence and universal pragmatics. In this context, consider the following at least plausible social-psychological hypothesis: "Personality stability in complex societies does not permit the individual to suspend his private interests, for such interests serve to integrate the many roles the individual must play." Were such a hypothesis correct it would pose a fundamental barrier to the realization of the ideal speech situation. It would invalidate Habermas' assumption regarding the potential generalizability of interests. It would show the decisive relevance of extra-linguistic factors.

The scientific study of language and its conditions, were the

---

arena of communicative action, and in particular to constitute the intersubjective object domain of the natural and social sciences in so far as these are subjected to the empirical method. The "objectivating sciences" are thus in the "Postscript," contrasted to the "self-reflective" sciences. ("Science and Objectivity," p. 100)

*Discourse is characterized by the "general symmetry requirements," which if adhered to would produce an ideally fair, rational, and legitimate consensus. The "general symmetry requirements" are that every participant have: (1) the same chance to initiate and perpetuate a discourse; (2) the same chance to put forward, call into question, justify, or refute statements, so that in the long run no statement is immune to criticism; (3) the same chance to express attitudes, feelings, and intentions, so that all may see if the participants are sincere and truthful; (4) the same chance to permit and forbid, so that the formal equality of chances to initiate discourse can actually be realized. These conditions are paraphrased from Thomas McCarthy's "A Theory of Communicative Competence," pp. 145–46.

above hypothesis correct (there is no reason to believe it is), would reveal that the context of language is, in a very real sense, not just language. Habermas could raise the fundamental objection that because science is itself a stunted language, it could not reveal the meaning of this fact without the assistance of the fully developed language whose limits it seeks to reveal. This meaning would be revealed in discourse, and its significance debated there. Yet, this response, while correct, hardly eliminates the challenge posed by the above hypothesis. Habermas does state that it is quite possible that all private interests may not be generalizable. However, he goes on to argue that this can only be determined in and through discourse. My argument is that other ways of determining limits on the generalizability of private interests cannot be a priori excluded, especially since these other ways might reveal the falsity of the discursively achieved consensus on generalizability.

Though Habermas rejects this possibility,[43] his argument presumes the issue in question: that the scientific discovery of constraints on discourse must presume the validity of discourse. However, one might ask in response why the scientific study of language could not discover the following: that the constraints upon discourse that deceive individuals into believing that they are suspending their private interests do not apply to small groups of professionals informally (i.e., not fulfilling the formal conditions of discourse) discussing abstract issues of truth and falsity. That such a discovery does not seem likely is hardly the issue. Habermas is raising a priori philosophical objections; I am posing in principle counterexamples. In a related fashion, Russell Keat argues, in *The Politics of Social Theory*, that "the primary rationale for Habermas' rejection of biologically conceived instincts is his desire to remove the 'objects' of psychoanalytic theory from the domain of empirical-analytic science which investigates causal relationships."[44] Keat is referring to Habermas' belief that the instincts can become utterly transparent to the reflective ego. Whether or not this is an accurate characterization of Habermas' position on psychoanalysis, it does seem to be an accurate characterization of why Habermas would deny that science could discover compelling constraints on discourse. Were discourse not removed from the domain of science, its role as arbiter regarding what may be validly scientifically studied would be subject to correction by science itself.

"*Zu Nietzsches Erkenntnistheorie*."—Most of Habermas' other arguments for the autonomy of language tend to be a version of those considered above, such as his division of disciplines concerned with language constructed so that "empirical pragmatics" (psycho-linguistics and socio-linguistics) cannot impinge on "universal pragmatics."[45] Yet, the basis of Habermas' assertion of the autonomy of language seems to lie deeper: in a theory of the relationship between prelinguistic and linguistic signs. It is Habermas' only genuine *explanation* of the scientific inexplicability of language. Though this theory has not been explicitly articulated by Habermas, its elements can be found in two of his essays, "Der Universalitätsanspruch der Hermeneutik" (The Claim of Hermeneutics to Universality), and "Zu Nietzsches Erkenntnistheorie (ein Nachwort)" (An Afterword to Nietzsche's Theory of Knowledge). Putting the elements together shows this explanation to be little more than a tautology.

In "Der Universalitätsanspruch der Hermeneutik," Habermas asks how language can be used to translate something nonlinguistic—science—into language.[46] In formulating his reply Habermas refers to Jean Piaget's concept of operative thought. Operative thought, which employs such categories as space, time, causality and substance, as well as certain formal-logical relations, develops individually earlier than language and is never really integrated with it. Language "sits on top of" this way of thinking. Language is capable of integrating in a genuine fashion [*eine echte Integration*] only those prelinguistic "paleosymbols" that are an expression of the attempt to give meaning to experience.[47] It cannot integrate those ways of thought that simply order the objects of experience in different ways. The universal claim of hermeneutics to understand the world finds its limit in the stunted language of science. The use of language for the organization of means-ends rational action and for the construction of scientific theories is not the employment of a genuine language.

> In this case natural language is released from the structure of intersubjectivity; language appears [*träte*] without its dialog constitutive elements and separated from communication, only under the conditions of operative intelligence.[48]

Once more, the claim that science apprehends nature only within the framework of the technical cognitive interest is used to limit the proper domain of science. In this case, the consequence is that science—its methods and its theories—cannot be translated directly into practice. Practice is the realm of rhetoric, and rhetoric can only be about the content of science: its actual technical achievements. It is this view that informs Habermas' position, discussed in the next chapter, that science may enter the world of everyday life only as technology.

"Zu Nietzsches Erkenntnistheorie (ein Nachwort)" puts these considerations in an interesting light. Habermas examines Nietzsche's claim that it is the metaphors of thought, which are apparently prelinguistic signs, that allow the "fantastic" manipulation of the world, though at first only in our imagination. Only after these metaphors are ordered by being integrated into the grammar of language can they be used by science to manipulate the world.[49] However, according to Habermas, Nietzsche believed that because these metaphors stem from our imagination they must remain fictions. Inserting them into the grammar of language orders them for the purposes of manipulation and control, but only insofar as they are made a prior condition of knowledge of reality. This must be the case, because while we can know that these metaphors serve the continued existence of men and women, we have no warrant to assume that truth somehow corresponds to that which serves human survival.[50] Habermas quotes Nietzsche: "Als ob die Wahrheit damit bewiesen würde, dass der Mensch bestehen bleibt!"[51]

For Nietzsche it appears that language does not "sit on top of" operative thought at all. Rather, a genuine integration takes place, one which suggests that the richness of the human imagination and scientific explanation are not alienated. Yet, Habermas argues, this integration contributes to the inadequacy of Nietzsche's scheme, characterized by its crude pragmatism and denial of the possibility of truth. Nietzsche's skepticism indicates that he could not break free of the ontological concept of truth. The "subjective conditions of possible objectivity" suggested to Nietzsche, says Habermas, that the concept of truth was itself a fiction.[52] Habermas suggests the following resolution of this "paradox." Under man's particular ability to symbolize, reality is objectified under the viewpoint of possible technical control. However, the success of instrumental operations in this viewpoint is not identical with the truth of empirical sentences from which the capability of technical control is derived.[53] We recall that Habermas is not referring here to objections against this view raised by the traditional philosophy of science, such as Popper's fallibilism. Rather, Habermas is referring to his own distinction between the validity of science and the objectivity of the natural world. One may refer to the experience of worldly objectivity as it is disclosed by a particular scientific theory as evidence. However, truth itself means—the concept of truth refers solely to—agreement under the general symmetry conditions of discourse. This is Habermas' consensus theory of truth.[54] Truth is nothing more and nothing less than agreement in discourse.

The adequacy of Habermas' consensus theory of truth cannot be addressed here. Hesse discusses its relevance for Habermas' theory of science in "Habermas' Consensus Theory of Truth."[55] The obvious objection to Habermas' consensus theory is stated by McCarthy in "A Theory of Communicative Competence." There McCarthy notes that Habermas seems to conflate an idealized sociological or historical account of how individuals come to *agree* on truth through a process of consensus building with the *concept* of truth itself.[56] The concept of truth has traditionally, even in the case of the instrumental theory of truth (the coherence theory of truth is an exception),[57] referred to the relationship between actual states of affairs and statements about them. Such obvious and traditional objections are, needless to say, hardly decisive.

The validity of science is thus defined by Habermas solely in terms of its efficacy as an instrument of prediction and control. The truth of any particular scientific claim depends solely upon and can be characterized entirely in terms of the agreement of scientists under discursive conditions. Habermas severs the link between discursive agreement upon truth and the world this truth is presumably about. Habermas avoids Nietzsche's skepticism, but only by disconnecting the concept of truth from the reality it is presumably about. The possibility of truth is rescued only by defining it in such a way that it refers solely to agreement among equals. Reinhart Maurer disagrees with this interpretation. He concludes that Habermas never rescues truth from Nietzsche's skepticism. Habermas simply embeds this skeptical view of truth in the species' interest in self-preservation.[58] Whether Maurer is correct or not, it does seem plausible that one reason Habermas defines truth as consensus is because such a concept of truth is as readily applicable to normative as to empirical claims. This is because the consensus theory defines truth merely as agreement under specified conditions. The correspondence concept of truth is less suitable in this regard (i.e., it doesn't seem to fit normative claims as readily as empirical ones) because it is not clear—unless one believes in Plato's forms or the like—what in the world normative claims would correspond to. If this interpretation is correct, Habermas relativizes the *concept* of empirical truth (which has traditionally referred to the relationship between statements about things and things) in order to put it on a par with normative truth. From this perspective, the point of Habermas' distinction between the validity of science and its objectivity would indeed be to salvage the objectivity of science from this incipient relativism.

The claim that language "sits on top of" operational thought is fundamental to Habermas' claims regarding the autonomy of language. It also informs his position, discussed in the next chapter, that the most important task advanced industrial societies face is the "translation" of scientific knowledge into a linguistic form in which it can become the property of communicating citizens. Yet, Habermas' claim that language "sits on top of" operational thought lacks independent backing; it is no more than a derivation from his interest theory. Furthermore, his assertion hangs strictly on definitions. He argues roughly in the form, "that which is pre-linguistic can later be integrated into language, that which is not cannot." One could more easily conclude the opposite. The categories of space, time, and substance are the fundamental categories of meaning upon which all understanding builds. Piaget's work is at least as amenable to this interpretation.[59] Finally, it should be noted that Habermas seems to bring many of the difficulties and complexities discussed above upon himself. Only a view that starts from a certain skepticism regarding the possibilty that science could truly know the world requires the separation of operational and linguistic intelligence in order to salvage the concept of truth, a concept that now resides solely within the realm of language.

CHAPTER 7

Epistemology or Politics?

HABERMAS' CRITIQUE of scientism is overwhelmingly epistemological. He demonstrates the necessary limits of science by linking it to the technical cognitive interest while establishing the epistemological autonomy of language. However, in some of his works originally published in the late 1960s, several of which are collected in *Toward a Rational Society*, Habermas combines this strategy with a political one. These are some of his most interesting essays. They are also among the most troubling, for they suggest that his epistemological strategy may have potentially undesirable political implications. This thesis is best developed by first turning to an epistemological claim: Habermas' assertion that science is in principle "monologic." However, before doing so a caveat is necessary. Habermas' political studies of science were written during the years he first developed cognitive interest theory. Since that time his studies of science have become considerably more subtle and sophisticated. Were he to write about science and politics today, presumably his work on this topic would be likewise. However, he has not written recently on this topic. Those practical political studies that Habermas has recently undertaken have been more concerned with systems theory (and functional thinking generally) than science. These more recent studies, as well as Habermas' distinction between science and systems theory, are examined in chapter 8.

In *Knowledge and Human Interests*, Habermas states:

> Deduction, induction, and abduction establish relations between statements that are in principle monologic. It is possible to think in syllogisms, but not to conduct a dialogue in them. . . . Insofar as the employment of symbols is constitutive for the behavioral system of instrumental action, the use of language involved is monologic.[1]

Habermas readily admits that the designation of science as "monologic" is an abstraction. The issue is whether it is a useful abstraction that captures something important about science. The following considerations suggest it is not:

1. According to Habermas, not even the most petrified (or advanced) version of the philosophy of science employs syllogistic thinking.
2. It is well recognized that the text or gloss of a theory (i.e., its linguistic expression, which often takes the form of an analogy, such as "electricity flows as though it were water in a pipe") contains a "surplus" of meaning which cannot be axiomatized.[2]
3. Most scientific theories are never axiomatized, and no empirical theory has ever been fully axiomatized.
4. Theories are not usually the most important influence on the progress of science in any case. More influential are discoveries, which frequently take the form of falsifiable models, such as the structure of DNA.
5. The claim that science progresses by "determinate negation" (p. 89), one syllogism "sublating" another, is simply too abstract and recalls Engels' "dialectics of nature."
6. Habermas recognizes as much when he asserts, in a footnote to "The Self-Reflection of the Natural Sciences," that "the prescientific basis of experience stored in ordinary language" is the source of the revisory power of abduction. The model of determinate negation simply explains the "inner logic" of the behavioral system of instrumental action.[3]

A somewhat different perspective is provided by Habermas' introduction to the English translation of *Theory and Practice*. Here the term "monologic" refers not just to syllogistic thinking but to contemplative philosophies that fail to reflect upon the social context in which they arise.[4] This use of the term "monologic" suggests that Habermas might consider it applicable to any philosophy but critical theory. Indeed, Habermas defines "monologic" about as broadly as critical theory has traditionally defined "positivism."* However, even though he uses the term "monologic" loosely—as technical argot and as rhetoric—Habermas' purpose is clear: to demonstrate that scientific information per se is necessarily irrelevant to social life.[5] In order for scientific information to become relevant the results of science must be *translated* into a form in which they can become the subject of discourse. The philosophical basis of this claim is discussed in the preceding chapter.

Habermas sometimes writes of "translation" as loosely as he writes of monologic science. Sometimes translation means merely that scientists and politicians talk with each other. Politicians tell scientists what they want and scientists tell politicians what is technically possible. Habermas calls this exchange "the dialectic of enlightened will and self-conscious ability."[6] The lives of politicians, suggests Habermas, are embedded in the web of linguistic interaction of everyday life, a web that embodies the richness of cultural traditions. Consequently, their decisions regarding the employment of scientific research can be seen as an almost automatic taking linguistic possession of science. Scientists and technical experts, on the other hand, are presumably not so attuned to the dogma and tradition of the society they live in. They are freer, therefore, to go beyond the boundaries of society's self-understanding and imagine new possibilities that stem from their expert knowledge. It must be noted, however, that though the participants may represent the general interest, this remains an elite dialogue. Participation is necessarily restricted to the highest levels.[7] As an instance of this "dialectic" Habermas cites discussions between high-level government bureaucrats who let research contracts and researchers at the universities and think tanks ("major research and consulting organizations").[8]

Though scientific knowledge is irrelevant to the lifeworld, ac-

*Thomas McCarthy defines "monologic" even more broadly. In his introduction to *The Theory of Communicative Action*, McCarthy states that the "monological approach preordained certain ways of posing the basic problems of thought and action: subject versus object, reason versus sense, reason versus desire, mind versus body, self versus other, and so on" (p. vii). To me it is questionable whether a concept that is extended to cover so much territory can remain a useful one.

cording to Habermas, technology is not. It is as technology that science enters the social world. At one level this claim certainly makes sense. Atomic weapons have had a far more profound impact upon society than have scientific debates about the structure of the atom. However, it should not be overlooked that scientific theories and ideas are often profoundly influential in their own right. The order and regularity of the Newtonian universe impressed several centuries and countless social theorists, and the theory of relativity has been used and abused to justify everything from existentialism to Christianity. It could be argued in reply that because Habermas sees science as an instance of the technical cognitive interest, even scientific theories come to be seen by him as a type of technology ("technical knowledge").[9] Thus the counterexamples of Newton and Einstein miss their mark. However, while it is indeed correct that Habermas defines technology broadly, the term in this context is primarily intended to encompass industry and bureaucracy rather than abstract scientific theories. Technique, states Habermas, is the scientifically rationalized disposal over objective processes in the form of a system in which research (here scientific theory is drawn into technology) and technology stand in a feedback relationship with economics and administration.[10]* Technique is no longer a tool to achieve particular ends. It is virtually coextensive with the boundaries of modern industrial society itself. This is why private education is no longer a medium by which scientific knowledge can be translated into practice via the decisions and discussion of enlightened men and women. The translation process has become embedded in the political structures of modern society.[11]

The Ambiguity of "Naturwüchsigkeit."—An ambiguity has slipped into Habermas' analysis. It is probably characteristic of any analysis

*Shapiro translates *Technik* as *technology*. Cf. "Technischer Fortschritt und soziale Lebenswelt," *Technik und Wissenschaft*, p. 113. *Technik*, in German, generally refers to the apparatus itself, whereas *Technologie* refers to means-ends rational action, which would seem to be a more appropriate term in this context. I conclude that the opposition between the technical and practical cognitive interests is so fundamental in Habermas' work that it swamps finer distinctions. See Habermas' "Praktische Folgen des wissenschaftlich-technischen Fortschritts" for a similarly extensive though somewhat more abstract definition of *Technik*, p. 337. Shapiro's "Translator's Preface" to *Toward a Rational Society* captures Habermas' expansive use of the term *Technik*, p. vii.

that interprets "technique" so broadly. It is certainly present in Marcuse's analysis. The ambiguity can be explicated by Habermas' use of the term *Naturwüchsigkeit*, which means literally growing-out-of-natureness.*

> The direction of technical progress is still largely determined today by social interests that arise spontaneously [*naturwüchsig*] out of the compulsion of the reproduction of social life, without being reflected upon and confronted with the declared political self-understanding of social groups. Consequently, new technical capacities break-through without preparation into existing life-forms.[12]

However, simply because interest groups and social classes influence the direction of technological progress hardly implies that science and technology per se have penetrated a lifeworld unprepared to cope with their intrusion. Habermas uses the existence of social conflict over the introduction of new technologies as evidence that "the reified models of the sciences 'migrate' into the socio-cultural life-world and gain objective power over the latter's understanding. The ideological nucleus of this consciousness is *the elimination of the distinction between the practical and the technical*."[13] In "The Classical Doctrine of Politics," Habermas employed the history of political theory as evidence to support claims made about the nature of science. Here Habermas uses conflict among "social interests" to support the claim that science and technology are ideologies that cause the species to forget how to act prudently. The distance between this claim and the form this forgetfulness takes, the conflict of interest groups or social classes, suggests that this issue might more fruitfully be approached in traditional political terms.

Habermas has not, to be sure, forgotten the importance of politics. Yet, too often the "breaking through of politics into technical decisions" is seen as applicable only to the fantasies of technocrats, such as Niklas Luhmann, and not to the claim of cognitive interest theory to have uncovered the key to modern ideology in its suppression of the distinction between science and rhetoric.[14] Habermas' tendency to conflate politics and epistemology has been re-

*Shapiro suggests this literal translation, as well as the more meaningful "unplanned, fortuitous development," in *Toward a Rational Society*, p. viii.

ferred to by critics seeking to explain why *Legitimation Crisis* does not fit together very well.[15] However, the clearest expression of this tendency is Habermas' earlier study of the peculiar status of discussions between politicians and scientists. Habermas labels these discussions "steering discourses." He recognizes that steering discourses must be embedded in the political process so that they may be more than a mere decoration. This is why he grants these discourses formal functions, such as formulating long-term research and development strategies, as well as the authority to grant research and development contracts.[16] At the same time, Habermas wants to invest these steering discourses with the hermeneutic interest of humankind as a whole in taking linguistic possession of science. Yet, government bureaucrats would seem to make poor transcendental agents of the practical cognitive interest, and it seems terribly optimistic to characterize their interaction with the employees of think tanks—with whom they are often professionally interchangeable in any case—as the "dialectic of enlightened will and self-conscious ability."

Habermas seems to have backed himself into a corner. Given his analysis of science and technology as an ideology, the last thing one should want as a solution is an elite group of bureaucrats and scientists making the key research and development decisions. William F. Buckley has claimed (the story is perhaps apocryphal) that he would rather be governed by the first several hundred people in the Boston telephone directory than the faculty of Harvard University. Fortunately, there are positions that fall between rule by an intellectual elite and rule by alphabetical lot. Though informed public dialogue over the goals science and technology should serve is perhaps currently impossible, as Habermas claims, elite dialogue is not necessarily the most desirable alternative.

It is not merely science that suffers from decisionism, suggests Habermas. Liberalism, understood as a political ideology shared by all Western democracies, also suffers from a form of decisionism in which it equates a fair and just social order with one that counts and adds the preferences of its citizens. The possiblity of the practical rationalization of these preferences, by which Habermas means the possibility that the individuals could learn to distinguish legitimate from illegitimate demands, including their own, is abandoned. Liberalism abandons the belief that citizens with conflicting interests could come together under "discursive conditions" and agree upon which interests genuinely partake of the public interest, and which are merely private (i.e., selfish). In its place liberalism has substituted a form of perpetual logrolling (vote trading), at least in prosperous times, in which most groups get at least a little of what they privately want, the amount being determined by the balance of political power. The relationship to scientific decisionism is clear. Both abandon belief in the possibility that preferences can be rationally justified. They become noncognitive choices, values, wants, or votes.

Critical theory's richest and most challenging analysis centers upon the way in which scientific and political decisionism arose together and reinforce each other. It is the central theme of Horkheimer and Adorno's *Dialectic of Enlightenment*, as well as Habermas' "The Classical Doctrine of Politics." However, Habermas' concern with overcoming decisionism seems to lead him to overlook more strictly political responses to the problem of rationalizing public policy-making. The term "rationalizing public policy-making" is employed here in Habermas' dual sense. It refers to possibilities for increasing genuine public participation in the decision-making process, as well as increasing the likelihood that the outcome will be an effective response to a complex environment.

A Partial Political Solution to an Epistemological Problem

In "The Scientization of Politics and Public Opinion," Habermas characterizes the political and technocratic models of decision-making. The clearest statement of the political model—Habermas calls it the "decisionistic" model—was formulated by Max Weber, who argues that politicians must choose the values they wish to achieve by an act of political will, often in the context of intense interest group conflict. The role of the expert is confined to providing means to realize these values. Expertise is subordinate to politics. Under this model, says Habermas, public participation is reduced to voting for or against competing elite groups who manage public affairs. In this case the elite groups are political, rather than technical.

> Decisions themselves, according to the decisionistic view, must remain basically beyond public discussion. The scientization of politics then automatically accords with the theory developed by Weber, extended by Schumpeter, and now unques-

> tioned by modern political sociology, a theory that in the last analysis reduces the process of democratic decision-making to a regulated acclamation procedure for elites alternately appointed to exercise power. In this way power, untouched in its irrational substance, can be legitimated but not rationalized.[17]

Here it seems that the "scientization of politics" means little more than mass democracy in the absence of widespread and informed public discussion of the issues.

It is somewhat surprising to find this argument regarding nonparticipation being applied to the decisionistic and technocratic models of decision-making in equal measure. Indeed, though Habermas intends to distinguish between these two models, it is not very clear how they differ. Under both, public participation is restricted to voting for elite groups that manage the affairs of society. However, exactly how, in what way, and to what degree technicians and bureaucrats must establish their ascendency over politicians in order that a particular regime may be called technocratic rather than decisionistic is not addressed. Under neither model, says Habermas, does the public do more than acclaim the right of various groups to lead, because under neither does it possess knowledge of the possible alternative futures. Such knowledge stems from an awareness of different technical possibilites, which in the decisionistic model remain in the hands of experts and politicians and in the technocratic model do not even leave the experts' hands. Indeed, Habermas' argument suggests that the main difference between the technocratic and decisionistic models is that the former lays claim to the technical rationality of its decisions (i.e., they are the most effective way to manage a complex environment), whereas the latter model cannot make even this claim.

In a brief article published over twenty years ago (1964), Habermas suggests that the outcome of political conflict may be regarded as legitimate to the degree that every individual affected by a particular policy is fully and effectively represented by an interest group during the formation of that policy.[18] Habermas sought, in other words, to make liberal pluralism live up to its own ideological claims. He has not, however, pursued this strategy, perhaps because it resembles the decisionism he so sharply criticizes. Even ideally fair political conflict seems unlikely to yield decisions that are somehow more than mere decisions. Instead, Habermas has focused most of his attention on highly restrictive political institutions, such as steering discourses, because only in such institutions is discursive justification possible. Only in such institutions can decisionism be overcome, because only under conditions of *actual* equality—of information, education, status, and self-confidence—can the general symmetry conditions of discourse be met. Indeed, these conditions are a definition of actual, not merely formal, equality. The "dialectic of enlightened will and self-conscious ability" has no room for the technically ignorant and politically naive.*

Several Institutional Innovations.—Several very simple institutional innovations consonant with Habermas' early (1964) position on legitimate political compromise are discussed below. These examples pertain to policy-making for issues with a major technical component to them, such as toxic waste disposal or the siting of nuclear power plants. If members of the public can be guided to a reasonable level of scientific sophistication regarding such complex issues, then it should be possible in other areas as well. These institutional innovations are not proposed as simple solutions to the complex problems Habermas and others have identified. By themselves they can make hardly a dent. The innovations are only intended to exemplify the types of solutions against which there appears to be a bias in Habermas' thinking about the problem. Habermas would surely not object to institutional innovations that would enable previously excluded individuals and groups to participate more effectively. Consider, however, whether there is any place for such innovations in Habermas' project. Consider too whether Habermas' failure to provide a place for such innovations is the result of a systematic bias, as I contend, and not simply a levels of analysis phenomenon.

One insight is especially important to those who would strengthen the ability of previously excluded groups to compete in the politics of policy-making for complex technical issues. It is that scientific knowledge is itself a scarce resource, much like any other. However, unlike many scarce resources it is not in principle scarce. My increase in scientific and technical knowledge need not be at the expense of yours, although it may cost you money for programs.

*Make no mistake, the elite character of the proposed steering discourses clearly concerns him. His goal remains the gradual extension of discourse to the entire society. See "The Scientization of Politics," pp. 74–75.

Establishing programs that would distribute this knowledge more widely would be a significant first step. Habermas is quite correct that "private education" in this area is no longer sufficient. Science and technology have entered the industrial system, and with it the political relations of society. The scientific and technical expertise of groups powerful enough to contest in this arena must be enhanced.

The National Science Foundation of the United States has had a program, "Science for Citizens," based upon the scarcity premise. The program includes support of "public service residencies" for scientists and engineers who want to design a project with a citizens group. It also includes "planning study awards," which would "enable scientists and citizens to plan and develop permanent scientific information and advice programs within the community."[19] The National Science Foundation is, however, a quasi-governmental group, and one should perhaps not expect the government to fund those who stand in fundamental disagreement with its policies. Indeed, such funding could lead to co-optation. However, several programs have been proposed and put into practice that require no governmental funds. For instance, the members of the Biophysical Society, a small and specialized professional group, possess the expertise necessary to determine the subtle biological effects of radiation and chemical contamination. Since about 1972 it has matched members in various fields with government and citizens groups seeking expertise.[20] The service and advice are free.

Such programs bring members of the informed public concerned with particular issues, but who lack the technical expertise to do such things as identify toxic substances, into contact with experts willing to help on either a short-term or a programmatic basis. Habermas' examples, on the contrary, are concerned with the relationship between government agencies and research organizations under contract to them. Making scientific expertise available to citizens groups may not perform a "steering" function per se. However, it may well contribute to the political rationality of public decisions by enlarging the number of effective participants. Furthermore, it can hardly harm the scientific rationality of the debate to have the opposition more rather than less scientifically informed. Indeed, as Habermas reminds us in regard to the technocratic position, those who would exclude participants from the decision-making process cannot demonstrate that they are not excluding more effective problem solutions than problems.[21] The actions of citizens groups in the United States, such as the Citizens League Against the Sonic Boom, have helped save us from several "technological Vietnams," in this case the Supersonic Transport.[22]

One of the most difficult problems the involved citizen or politician faces is comprehending the nature and extent of scientific uncertainty. Fully certified experts disagree fundamentally, often using data that differ by orders of magnitude. How is the citizen to judge between Harvard Professor *X* who says it's all perfectly safe, and Yale Professor *Y*, who says it's deadly dangerous? Treating science as a product to be given to citizens in the form of a menu of alternative futures can only render the citizen confused and alientated when, as is usually the case, experts disagree over the viability of various futures. (In "On Social Identity" [1974], Habermas recognizes this point in principle, although he does not pursue it.)[23] What most urgently requires translation is the process of scientific inquiry itself, so that citizens may come to understand why scientists disagree. Ignorance of the fallibility of scientific inquiry leads to intemperate expectations of science.

Ignorance of scientific inquiry can in part be overcome by private education, by which Habermas means the formal education of individuals in schools. However, Habermas is quite correct that this is insufficient. As more and more public choices are characterized by expert disagreement, it is also necessary to translate the process by which an expert, such as a nuclear engineer, comes to take a political stand on an issue such as the safety of nuclear power. There is no better way to learn how facts and values melt together in such cases than by participating in an issue-oriented citizens group, which must confront the claims of the opposition's experts with the counterclaims of one's own. Large sectors of the public in the industrial democracies currently possess sufficient education, free time, and political will to participate in such citizens groups were the political opportunities available. This last phrase may appear to be an enormous caveat, but in the present context it is not. The present context is one of intellectual strategies, not concrete proposals for reform. Once it is recognized that public policy can be rationalized, in Habermas' dual sense, by means that enhance public participation rather than restrict it, a great deal of traditional political analysis geared toward enhancing institutional responsiveness and transparency becomes relevant.[24]

Whether such citizens groups, composed of experts and laypersons, could contain scientism is an empirical question, though it is one currently quite impossible to test. While there have been successes, the depth and breadth of Habermas' criticism suggests that we can hardly regard the effectiveness of the Citizens League Against the Sonic Boom as evidence that the political containment of science is possible. Were citizens groups to become more influential in the making of public policy, perhaps they too would foment a series of "technological Vietnams," possibly because they could conceive of no alternative to "bigger and more is better." Were this the case, then Habermas' thesis about the species' forgetfulness of the distinction between science and rhetoric could hardly be ignored. Habermas could, after all, be correct. However, while it is prudent to retain an open mind about this possibility, it is hardly prudent to prejudice the answer in advance, as Habermas' analysis tends to do. That is, Habermas' approach could lead one to ignore or downplay genuine opportunities to expand public involvement on the unproven assumption that the public is incapable of distinguishing between science and rhetoric.

One could argue that Habermas is merely strategically mistaken. The rational public discussion of the goals science and technology should serve does indeed lack an appropriate rhetorical medium. We no longer know how to discuss social goals rationally and prudently. Not even citizens groups can avoid the profound, though often subtle, effects of the decline of the classical doctrine of politics over two millennia. Nevertheless, one could continue to argue, the appropriate short-term strategy must be to increase opportunities for public participation, rather than fostering elite steering discourses. Otherwise the conditions of rational public participation will never be realized. These conditions can only be built from the bottom up, not from the top down. This requires that considerable resources be devoted to enhancing the conditions of public participation over issues that are perhaps quite secondary, such as where nuclear reactors should be built. Questions regarding, for instance, the relationship between the good life and "hard" and "soft" energy paths come later.

Such an argument would let Habermas off the hook only by not taking him seriously. His claim that modern science and technology have, via the ideology of scientism, caused the entire species to forget the distinction between instrumental action and rhetoric is nothing less than a political version of the dialectic of Enlightenment. Like Marcuse's concept of the technological totality, the concept of scientism occasionally seems to substitute in Habermas' analysis for what would really be a more appropriate explanatory strategy: the detailed social, economic, and political analysis of the conditions that too often lead to the exclusion of the public from all but symbolic acts of participation. Couple this substitution with a philosophical preference for justification over political compromise, and the source of a certain exclusionary or elitest tendency in Habermas' early political studies of science—a tendency that works against Habermas' intent, as well as his genuine commitment to the value of free and open participation—becomes clear.

CHAPTER 8

Habermas' New Science

HABERMAS' NEW science seeks to realize the goal of "Wozu noch Philosophie?": a non-scientistic philosophy of science. Reconstructive science attempts to overcome the conventional origin of critical theory's concepts: that they have emerged within and depend upon a particular constellation of historical, social, and psychological factors. The reconstruction of cognitive competencies seeks to explicate the basic cognitive strategies that individuals employ to make sense of their world. Reconstructions are claimed to be prior to science because they are about the basic cognitive competencies, such as those outlined by Piaget, that make science possible. Habermas first claimed that reconstructive science was not falsifiable by ordinary science. Under criticism he has backed off from this claim, but precisely how reconstructive science differs from ordinary science has become less clear.

From Science to Luhmann's Systems Theory

Habermas' changing approach to science is most clearly seen in his encounter with Niklas Luhmann's systems theory. (The focus here is exclusively upon Luhmann's systems theory. Habermas' extensive discussion of Talcott Parsons, in *Theorie des kommunikativen Handelns*, is not as relevant to the issue of scientism. Aspects of Habermas' treatment of Parson's systems theory are considered later in this chapter.) In this encounter science becomes an ally of critical theory in deflating the pseudo-scientific claims of Luhmann's systems theory. Indeed, the focus of systems theory upon survival—its de facto elevation of the survival of a particular system into its paramount goal—suggests that systems theory all along was a more appropriate target than science.

In "Dogmatism, Reason, and Decision: On Theory and Praxis in Our Scientific Civilization," Habermas lists the four stages by which technology itself becomes a value system:

1. "We employ techniques placed at our disposal by science for the realization of goals." At this level the rational control of nature depends solely upon the level of development of science and technology.
2. "If, however, there is a choice between actions of equal technical appropriateness, a rationalization on the *second* level is required. . . . [T]he information furnished by empirical science is not sufficient for rational choice between means which are functionally equivalent." This is the level of decision theory, which "analyzes the possible decisions normatively in accordance with a rationality of choice defined as 'economical' or 'efficient.'"
3. Values themselves are changed so that they may be readily operationalized. "The originally invested values, that is, those value systems with which decision theory is initially solely occupied, are then relativized in terms of . . . strategic value, by which the game or the conflict is given its orientation." An instance of strategic value would be the transformation of states of the world into numbers.
4. The goal of politics becomes transformed into mere system stability and survival. "Then ultimately it is sufficient to reduce all value systems to an, as it were, biological basic value, and to pose the problem of decision-making generally, in the following form: How must the systems by which decisions are made—whether by individuals or groups, specific institutions or entire societies—be organized in order to meet the basic value of survival in a given situation and to avoid risks."[1]

The impression is conveyed by this scheme that strategic action is a heightened form of scientism. At the first two levels normative questions are excluded from scientific argument; at the latter two levels the basic values of technological social planning take their place. However, the relationship of scientism and strategic action in Habermas' work is more complex and contradictory.* This point is best pursued by first noting that Luhmann's systems theory is characterized by its abstract comprehensiveness. It is not merely an empirical systems theory like biocybernetics, for example, but is instead virtually a metaphysic. It seeks to explain the development of society as a whole, including such subsystems as science, as well as the development of systems theory itself, in terms of the basic system theoretic categories of complexity and complexity reduction. The details of Luhmann's scheme are, however, not important here. What is important is that Luhmann's systems theory makes claims that are on the level of generality of Habermas' cognitive interest theory. As such, it is a philosophy, or as Habermas puts it, a claim to have reconstructed the rules that all societies employ to solve the fundamental problem of existence: the reduction of complexity.[2]

As a philosophy, Habermas considers systems theory a bad advance over scientism. Scientism was never able to justify practical questions. For this reason it led to decisionism and the consequent remythification ("demons") of the practical realm. Systems theory, though, represents the highest form of technocratic consciousness, for it defines all normative questions as technical ones and so removes them from public discussion.[3] The complexity of the environment, in Luhmann's view, requires a high degree of administrative autonomy. The demands of world complexity place such great pressures on the adaptive capabilities of society that society

*In terms of Habermas' system, strategic action falls together with science under the rubric "instrumental action." Each is monologic, by which Habermas means here that each uses language to better manipulate the environment, rather than as the medium of mutual recognition. In fact, it is not precisely clear whether Luhmann's systems theory falls neatly within the category of strategic action. Its abstract comprehensiveness, discussed below, seems to put it "above" this category, yet it seems to fit stage four of rationalization quite precisely. To add to this complexity, Habermas' theory of social evolution grants systems theory its own path or mode of evolution, distinct from the evolution of instrumental action. Perhaps it is most accurate to say that Luhmann's systems theory is an instance of the technical cognitive interest applied to society. In any case, Luhmann's systems theory stands closer to the technical than the practical cognitive interest.

can ill afford to let policy be the outcome of political compromise among competing groups. Such policy is bound to reflect the balance of power among groups rather than being the most efficient response to a problem-posing environment. Rational social policy thus requires comprehensive nonparticipatory social planning by an administration protected from party politics and participatory demands by the public. Consequently, it requires that its decisions be regarded as legitimate solely because they are formally legal. Questions from the public regarding the normative quality of these decisions interfere with the rationality of the output.[4] Habermas states that Luhmann's systems theory is the realization of his worst fears regarding the obliteration of the distinction between the technical and practical.[5]

This bad advance from science to systems theory involves a jump from one type of reasoning to another rather than being a continuation of scientism by other means, as the four stages might seem to imply. Although both scientism and systems theory are in principle monologic—and thus part of instrumental action—Habermas now grants systems theory its own separate mode of social evolution: science is an expression of cumulative learning within the framework of instrumental action, whereas systems theory is an expression of enhanced steering capacity within the framework of heightened social complexity.[6] Because science and systems theory reflect quite different cognitive strategies, science can now act as an ally of critical theory against the epistemological imperialism of systems theory. Science ideally makes demands upon itself, such as that its theories be falsifiable, that systems theory cannot meet. On these grounds Luhmann's systems theory can be convicted as pseudo-science.

> Scientism also sets standards by which it can itself be criticized and convicted of residual dogmatism. Theories of technocracy and of elites, which assert the necessity of institutionalized civil privatism, are not immune to objections, because they too must claim to be theories.[7]

Not science, but systems theory, it now appears has inherited the legacy of Machiavelli and More. Systems theory expresses the primacy of survival. Science, far from being the paradigm of such thought, now upholds the Enlightenment ideal of a concept of truth that transcends functional utility.[8] Actually, Habermas has held such a view of *truth* all along. He simply disconnected it, via discourse, from the instrumental activities of science. With Luhmann, Habermas has found a willing target: someone who is eager to acknowledge that all knowledge is a type of social technology in the service of survival in a complex world.

The Reconstruction of Cognitive Competencies

If cognitive interest theory, as Habermas once argued, possesses a quasi-transcendental status, then one might say that reconstructions possess a quasi-scientific status. Reconstructions are the methodology of the theory of social evolution. Like cognitive interest theory, the theory of social evolution seeks to demonstrate that societies develop along two key dimensions, the instrumental and the practical, that are irreducible to each other. Like cognitive interest theory, the theory of social evolution is a reconstruction of historical materialism. Both cognitive interest theory and the theory of social evolution seek not only to document the autonomy of the linguistic "superstructure" but also to establish that it is innovation in culture and language rather than in science, technology, and industry, that is the real pacemaker of social development.[9] The terms change but the strategy remains similar. However, unlike cognitive interest theory, the theory of social evolution avoids the difficult position of claiming that human knowledge is at once grounded in the world and constitutive of it. Yet, the quasi-scientific status of reconstructions is ultimately as problematic as the quasi-transcendental status of cognitive interests. It will be suggested that Habermas' reluctance to treat reconstructions as straightforward scientific theories or hypotheses stems in large measure from his employment of them as an empirical foundation for a nonempirical claim: that the telos of social evolution is toward progressively less constrained dialogue.[10]

The Theory of Social Evolution.—Since the concern here is not with the theory of social evolution per se but rather with its reconstructive methodology, I shall simply outline the former.[11] The theory of social evolution assumes that it is possible to reconstruct the pattern of development of the cognitive and especially the normative structures of society.[12] The developmental stages that emerge from this study can be seen as embodying or representing rules for possible problem solutions. However, in order to clarify this developmental learning process, an empirical theory of how individual learning becomes available to society is needed. In this regard, Habermas states that we should look for the actual learning mechanisms only at the level of the individual. Societies learn only in a metaphorical sense.[13] Habermas finds this learning mechanism in social movements, which frequently begin with the unique insights (or world views) of one or a few individuals. Social movements constitute a major vehicle by which individual learning is made available to society.

It is apparent that Habermas is concerned with two types of learning: learning that is a response to ordinary challenges to the system, and learning that represents a unique response to an exceptional challenge. It is the latter type of learning that lifts society from one developmental level to another. Insofar as is possible, says Habermas, system problems are solved within the established learning level. The established learning level is the society's organization principle, which "may be characterized by the institutional core which determines the particular dominant form of social integration."[14] However, some system problems are so fundamental that they cannot be solved in accordance with the world views characteristic of the existing learning level. In these circumstances, says Habermas, if the society is sufficiently culturally rich and differentiated, there may be groups or individuals within it that hold world views relevant to the solution of the system crisis. If the society proves capable of drawing upon these world views (this is, of course, how individual learning is brought into the service of the social system), it may solve its system crisis. However, it can do so only by abandoning its existing learning level, and thus its existing form of institutional integration, for a new one, which incorporates the successful world view. It is by this process that societies move from one developmental level to another. "Societies can learn in an evolutionary sense by using the cognitive potential stored in worldviews for the reorganization of action systems. . . . Introduction of a new organization principle means the establishment of a new level of social integration."[15]

Habermas' scheme does not posit that societies must evolve. Indeed, what might be called a permanent regression is possible. Nor does his system posit that this evolution must follow a particular sequence. Some societies might skip or repeat some stages.[16] Habermas thus avoids the most problematic aspects of evolutionary theories, aspects which Karl Popper has termed "historicism."[17] What Habermas' scheme does posit is that at a particular level of development a society will have a particular repertoire of responses available to it, characterized by the achieved level of learning, and thus the achieved level of social integration. While the sequence of stages itself is not fixed, a society will not be able to call upon responses characteristic of developmental level *A* to solve problems it faces while at developmental level *B*. It is this assumption that gives Habermas' theory its empirical content.[18] Habermas also notes that the relationship between individual and social development remains only "homologous." He recognizes that he has not identified a mechanism but only a link, which mediates individual and social evolution.[19] Nevertheless, the idea is clearly that the sequence of stages of individual moral development explicated by Lawrence Kohlberg and others (as well as the sequence of stages of individual cognitive development explicated by Jean Piaget), can serve as a model for the logic of social evolution. For example, the stages of individual moral development serve as a model to indicate the repertoire of responses available at a particular level of social integration. Social development is modeled after personal development.

Habermas' theory of social evolution has not escaped criticism. Michael Schmid argues that the claimed homologous relationship between individual and social evolution is part of a sub rosa attempt to derive ought from is. Schmid argues that while Habermas talks in terms of the "evolution," "development," or "progress" of societies from pre- to postconventional principles of organization, he has no grounds to use any of these terms or their cognates. All that may be justifiably said about different principles of organization from an empirical perspective is that they are different.[20] However, Schmid continues, what Habermas actually does is to imply that because postconventional morality can serve as an organizing principle around which to solve certain system crises, it is somehow more ethically advanced. In fact, says Schmid, there are many societies in which the introduction of postconventional norms would seriously endanger social integration.[21] On what grounds are we to say that these societies are ethically less evolved or developed? Because they do not correspond in a homologous fashion to the postconventional stage of individual moral development? This could hardly be the answer because it begs the question at stake: why is postconventional morality better and not simply different?

Why does Habermas come so terribly close to committing the

naturalistic fallacy? The answer, says Schmid, lies hidden at the conclusion of Habermas' "Geschichte und Evolution."[22] Schmid argues that there Habermas suggests that his evolutionary theory is to be applied in discourse, where competing definitions of the good life are at stake. Discourse, it will be recalled, is designed to achieve rational consensus over such issues. However, without standards such consensus risks being mere consensus; i.e., arbitrary. The theory of social evolution, though, provides a standard, insofar as it says, in effect, that some solutions or policies are better than others, because they tap the achieved level of morality better than others. In other words, because every stage of social integration (development) has a particular repertoire of moral responses associated with it, this level can serve as a standard by which to rationally reject responses characteristic of lower levels. However, once it is recognized that the repertoires of responses are merely different, not higher or lower, the utility of the theory of social evolution as an implicit moral standard in discourse evaporates. This is why, says Schmid, that Habermas has not directly confronted this issue.[23]

Schmid's is the most comprehensive, and among the harshest, criticism of the theory of social evolution. However, he is not alone in detecting a certain eagerness on Habermas' part to downplay fine distinctions between is and ought. Thomas McCarthy, generally among the most sympathetic critics, is remarkably stringent in his criticism on this point. He notes, for example, that anyone familiar with the "rationality debates" that have accompanied the development of cultural anthropology would find it far from obvious that the ability to reason argumentatively and reflectively is "the realisation and completion of potentialities which are universal to humankind."[24] McCarthy also notes that the charges of ethnocentrism and rationalistic bias that have been leveled against Piaget and Kohlberg—theorists whose schemes are models of an evolutionary reconstructive science for Habermas—have not been adequately addressed by Habermas.[25] As an example of the problem raised by developmental models of morality, McCarthy notes that whether the transition from "contractarianism and utilitarianism to 'justice as fairness'" is an advance to a higher stage (and not regression or mere change) is a question for meta-ethical argument, not empirical study.[26] Though Habermas has been quite responsive to these criticisms (which are, it should not be overlooked, directed against Piaget and Kohlberg, not merely Habermas), a related set of criticisms can be brought against Habermas' reconstructive science. These criticisms are related because reconstructive science (the methodology of the theory of social evolution) itself also seems designed to find support for ethical claims in a special science of nature.

Reconstructive Science

Habermas employs Noam Chomsky's analysis of linguistic competence as paradigmatic of a reconstructive science. Other paradigms are Jean Piaget's genetic epistemology and Lawrence Kohlberg's explication of the stages of moral development. Linguistic competence, says Habermas, is not an abstract theoretical concept, but a "know-how" that expresses an actual underlying structure in language, which can be reconstructed by eliciting the speakers' intuitions. The relationship between linguistic theory and linguistic data thus differs from the relationship between theory and data in ordinary science. This is because ordinary language is definitive for linguistic theory. That is, linguistic theory, understood as a reconstructive science, is nothing but the explication of the rules of ordinary language use. Thus, Habermas claims that "a proposal for reconstruction . . . can represent pretheoretical knowledge more or less explicitly or adequately, but it can never falsify it."[27] It can never falsify this pretheoretical knowledge, but only reproduce it, because reconstructions are descriptive of this ordinary use. They do not impose a theoretical structure upon it.[28] Reconstructions, says Habermas, "make an essentialist claim. . . . [I]f they are true, they have to correspond precisely to the rules that are operatively effective in the object domain—that is, to the rules that actually determine the production of surface structures."[29]

Reconstructions, in this view, are templates. They are descriptively accurate accounts of cognitive processes. It is the cognitive processes described that are nomological and universal, in the sense that they act to make the world appear to us in this fashion. Reconstructions are not, like ordinary science, a theoretical structure imposed on the data.[30] Rather, reconstructions describe actually existing theoretical entities: the rules of ordinary language use in this case. The theories of a reconstructive science are realistic; they capture the essence of actual entities.[31]

The essentialism of reconstructions overcomes the conventional origins of concepts, according to Habermas. While social theory cannot help but reflect the historical circumstances within which it arises, reconstructions go behind these circumstances to capture their conceptual ground. Reconstructions overcome the situational relativity of knowledge claims.[32] They are interest free, because they are prior to and a condition of cognitive interests, as well as more transient social, economic, and political interests. Reconstructions are empirical descriptions of core cognitive processes. However, because of their essentialist status they can hardly be shown false by an alternative empirical hypothesis, because reconstructions are about the cognitive prerequisites (language and the protoscientific cognitive abilities that apprehend nature under the categories of space, time, and substance) of any such hypothesis and its evaluation. Reconstructions are scientific, but not falsifiable by scientific evidence. Habermas once made a similar claim about language. Language cannot be fully scientifically explained because to do so one would have to employ the linguistic capabilities that one set out to explain in the first place. Such an argument, it will be recalled, identifies the scientific explanation of language with its replacement by science (the "displacement hypothesis"). Doing so frames the issue in the most extreme terms.

Habermas' reconstructive science has been subject to criticism at least as extensive, and perhaps more so, than that directed against his cognitive interest theory.[33] Perhaps this is because cognitive interest theory quite arguably makes a transcendental claim, whereas reconstructions seem so obviously an instance of ordinary empirical science that it is difficult to see the justification for treating them as a special category of science. Some of this criticism of reconstructions has already been mentioned. If the work of Piaget and Kohlberg epitomizes a reconstructive science, and yet their work is terribly susceptible to the charge that it is culturally biased, then the claim that reconstructive science is so fundamental that it bypasses cultural influences and the like is obviously vulnerable.[34] At a more general level, Mary Hesse notes that scientific and philosophical reconstructions of concepts such as human freedom and responsibility have had a profound effect upon the interpretations put upon ordinary discourse about guilt, punishment, deviance, and sickness. One has, she notes, only to think of changes in attitudes toward sexual "deviation," juvenile delinquency, and "diminished responsibility" in general.[35] It is thus hard to see how Habermas could claim that reconstructions are so basic that they define but (unlike more abstract ordinary theories) do not feed back to alter fundamental concepts of ordinary discourse. Habermas could respond, of course, that the reconstructions he has in mind are about more fundamental and universal processes than those referred to by Hesse. It is not clear, however, what standards Habermas would apply to support this claim. In any case, presumably the standards by which Habermas would argue that the reconstructions of Piaget are more fundamental than those of Freud, for instance (whose reconstructions have certainly influenced ordinary discourse), would themselves be hypothetical standards and thus open to question.

There is indeed much that is puzzling about the status of reconstructions. If the explanations of ordinary science are bound to the framework of the technical cognitive interest insofar as how they may be tested is concerned, this would seem to give these interpretations a certain stability. This seemed to be Habermas' point in upholding the embeddedness of science in the technical cognitive interest against the incommensurability asserted by theoretical pluralism. Why it is that reconstructions, which are not bound to the framework of the technical cognitive interest, would be more stable than the theories of ordinary science is not obvious. Apparently Habermas believes in the possibility of pure descriptions. However, even Habermas has suggested that Chomsky is mistaken if he thinks that universalistic theories of linguistic competence can be ideologically neutral.[36]

Perhaps the best way to resolve some of these difficulties is simply to note that Habermas has backed away from some of the most ambitious claims he had made for the special status of reconstructions. In this regard I quote his response in full to Hesse's criticism of the claimed special status of reconstructions.

> If I understand her correctly, Hesse's critique of my distinction between nomological and reconstructive sciences is directed against assertions that I did not put forward in the sense in question. The essentialism which, in my view, attaches to reconstructions of pretheoretical knowledge of competently knowing, speaking and acting subjects is not meant to deny that we are dealing here with fallible hypotheses, just as in the objectivating sciences. However, in the attempt to transform implicit abilities into explicit knowledge, the *terminus a quo* is connected with the *terminus ad quem* internally and

> thus in a *different* way than an existing state of affairs is connected with its theoretical description when the latter is based on a nomological theory.[37]

Leaving aside the issue of the accuracy of Hesse's critique, it is difficult to know what to make of this response. Habermas seems to agree that reconstructions are fallible hypotheses, "just as in the objectivating sciences." However, one must ask what it means to say that the *end from which* (i.e., point of origin) is connected with the *end toward which* (i.e., conclusion) differently in the case of reconstructions than in ordinary science. Presumably it means that reconstructions describe pretheoretical know-how, whereas ordinary science explains it under a theoretical structure. However, this is precisely the issue in question. In admitting that reconstructions are "fallible hypotheses, just as in the empirical sciences," Habermas would seem to have abandoned any basis from which to claim that reconstructions are more descriptive than the theories of ordinary science. Here we see what is really a pattern in Habermas' work, a pattern that is at once admirable and troublesome. (1) Habermas formulates special categories of knowledge, such as cognitive interest theory, or reconstructions. (2) He is extraordinarily responsive to criticism, revising and modifying his original categories, often extensively. The introduction of discourse is a prime example. (3) However, he continues to make claims or draw conclusions—now frequently in a highly qualified or implicit fashion—which are supported only by the original distinctions, which Habermas has abandoned under criticism. In the immediate case, once it is agreed that reconstructions are falsifiable generalizations about cognitive processes, it is difficult to see how they differ from ordinary scientific hypotheses in any relevant or important respect.

In *Theorie*, Habermas states that "if one is still willing today to venture to expound the universality of the concept of communicative rationality, without falling back upon the guarantees of the great philosophical tradition, basically three ways present themselves."[38] A consideration of these three ways is relevant to the issue of the status of reconstructions. The first way of exploring communicative rationality, says Habermas, would be to construct hypotheses regarding which patterns of communicative rationality are in fact universal and to check these against the actual intuitions and practices of speakers in a wide variety of societies and cultures. The second way would be to employ the theory of communicative rationality (i.e., universal pragmatics) as a practical technique, for example in the diagnosis of pathological communication, in order to check its empirical effectiveness and relevance. The third way, says Habermas, is to employ the theory of communicative rationality to interpret and reconstruct the tradition of social theory that runs from Weber to Parsons.[39] Habermas, of course, takes the third way. He characterizes it as "somewhat less demanding," but surely he is being modest.[40] It is difficult to imagine a more demanding intellectual task than the reconstruction of contemporary social theory. Nevertheless, it might have been advantageous were Habermas to have paid more attention to the first way of validating his claims about the universality of certain patterns of communicative rationality. Though the empirical evidence gained from cross-cultural studies and the like would almost surely be inconclusive at this point, such an approach might have helped remove the issue of testing reconstructions from the stratosphere of debates over fine epistemological distinctions. A more empirical approach to communicative rationality might have given the discussion of the status of its methodology—reconstructive science—a more concrete tone. The debate that followed would then have presumably also been about the validity of apparent exceptions to Habermas' hypotheses and so forth rather than being about what can only be called transcendental distinctions between different types of scientific knowledge.

Thomas McCarthy states that reconstructions represent "the scientific counterparts" to the transcendental and developmental perspectives in classical German philosophy.[41] Putting it this way helps to clarify the actual intent of reconstructions in Habermas' work. This intent is the same as that expressed by Habermas in "Wozu noch Philosophie?" There, it will be recalled, Habermas stated that contemporary philosophy should take the form of a "nonscientistic philosophy of science."[42] A reconstructive science is such an enterprise. As originally formulated by Habermas, it takes into account—indeed, is based upon—some of the most advanced empirical studies of human development. However, due to its essentialist status it is not falsifiable by ordinary science. Its truths are thus not contingent upon the advance of science. Perhaps they are as permanent as human nature. Nor in this interpretation are reconstructions unavoidably tainted by ideology, culture, and history. The possibility exists that a level of experience beyond these influences can be tapped by reconstructions. Furthermore, because the reconstructive science that taps this level of experience is not bound by the technical cognitive interest, it need not dehumanize its subjects by treating them as functional equivalents of bodies in motion. With reconstructions, a humane science of humanity becomes possible for the first time, because reconstructions grasp the actual cognitive processes themselves, not merely that aspect of them subject to technical control.

Habermas may well be correct regarding many of these claims. A science that could penetrate culture and ideology to apprehend humans as they truly are and not dehumanize them in the process is attractive, not absurd. However, if the distinction between ordinary science and reconstructive science is not categorical, but at most only a matter of degree, then much of Habermas' project—the containment of ordinary science, its confinement to the explanation of nature, so that science would not dehumanize humanity—has been misdirected. The very idea of a reconstructive science, coupled with the fact that it is virtually indistinguishable from ordinary science, suggests that it is not science per se that must dehumanize its subjects. This danger is rather posed by particular scientific theories or approaches, such as Luhmann's systems theory, naive social physics (e.g., Hobbes), or functional thinking generally. I have certainly not shown that ordinary science can explain human behavior from a perspective not controlled by the technical cognitive interest. It is Habermas' own reconstructive science, as modified to meet the objections of Hesse and others, that has done so. It has done so *de facto*, if not *de jure*, if I may put it that way.

What Purpose is Served by Habermas' Distinction Between Ordinary and Reconstructive Science?

Perhaps the reason that Habermas draws such fine distinctions between types of science is because Habermas recognizes that he risks doing what he has warned against all along, particularly in his critique of "self-objectivation." That is, Habermas seems to be explaining such fundamental aspects of meaningful human action as moral choice in terms of the operation of natural processes that can be scientifically studied. In order to be consistent, Habermas could reject his earlier criticisms of the intrusion of science into the communicatively structured lifeworld, i.e., his critique of scientism. He does not do so for several reasons, the most important of which is apparently that he does not want to open the door to *any* scientific explanation of human action. If the scientific explanations of Piaget and Kohlberg cannot be systematically distinguished from those scientific explanations that treat human beings as things ("self-objectivation"), then the theory of social evolution would itself seem to be an instance of scientism.

The primary way in which Habermas seeks to perform this gate-keeping function is to argue that the work of Piaget and Kohlberg exemplifies a special category of science that is more descriptive than ordinary science. Their reconstructive science does not impose a theoretical structure upon human action that is alien to human experience. Rather, Piaget and Kohlberg systematize concepts which have a basis in everyday experience (e.g., individuals' moral beliefs frequently become more abstract as they mature). An explanation of human action in terms of genetics or biochemistry, on the other hand, would employ terms and categories that find no basis in normal human experience. However, Habermas' distinction (while an accurate characterization of how scientific explanations that refer to unfamiliar entities are frequently experienced) lacks any basis in the epistemic characteristics that distinguish ordinary from reconstructive science. This is especially the case once it is recognized that the "displacement hypothesis" is a phantom, i.e., it is not a question of a neuro-physiological explanation, for example, replacing an explanation that employs more familiar categories. Each may be employed to explain different aspects of the same phenomenon.

Habermas seeks to translate his preference for explanations that are based upon familiar concepts—a preference that may have a valid basis in criteria such as the degree of subjective satisfaction that such explanations afford—into epistemological categories. It may be that concepts such as familiarity and subjective satisfaction could be worked up into useful epistemic criteria. However, the distinction between ordinary and reconstructive science provides few hints as to how this may be accomplished. It may also be that valid ethical and moral objections against explanations that invoke totally unfamiliar scientific categories can be raised, e.g., such explanations treat people as morally equivalent to things rather than as infinitely valuable ends-in-themselves. However, once more it is the case that the distinction between ordinary and reconstuctive science provides little leverage over this issue. One reason this is so is because Habermas' distinction between ordinary and reconstructive science is at base a distinction between description and explanation. Such a distinction, even were it to have a valid basis, does not seem to be a fruitful way to distinguish those scientific approaches that objectify human action from those that do not. Many explanations that are morally objectionable because they objectify people employ quite familiar categories. Aristotle, for example, compared some people with tools found in the household and workshop in order to justify slavery (*Politics*, bk. 1, chap. 4). What Habermas seems to want to say is that Piaget and Kohlberg capture the right aspects of human nature, i.e., those that are truly operative and relevant for a theory of social evolution. He may be correct. However, this is a contingent empirical claim and should be treated as such. To grant this claim a special epistemological and methodological status in order to protect it or to distinguish it from ordinary scientistic studies is jejune.

Theorie des kommunikativen Handelns

Toward the conclusion of the second volume of *Theorie des kommunikativen Handelns*, Habermas once more asks what the critique of positivism in a postpositivistic era should look like.[43] His answer in *Theorie* suggests that such a critique has already been adequately undertaken by the so-called "new philosophy of science" of Thomas Kuhn, Paul Feyerabend, Imre Lakatos, Mary Hesse, et al.[44] Today, he suggests, the critique of instrumental reason should focus strictly upon the reification of social relations, not upon man's relationship to nature. Indeed, this has been Habermas' position all along. The critique of instrumental reason need no longer focus upon science per se for two reasons. First, the self-understanding of science exemplified by the new philosophy of science is so modest, so temperate in its claim to exemplify reason, that it is no longer a threat to reason. Indeed, says Habermas, the "admirably consistent self-criticism" of the new philosophy of science has led it to understand itself as virtually a hermeneutic discipline.[45] Because the new philosophy of science grasps that data cannot be analyzed independently of the theories that constitute the data, it understands that science can best be understood as proceeding in an interpretive circle. Habermas quotes Mary Hesse approvingly: "The language of science is irreducibly metaphorical and unformalizable, and the logic of science is circular interpretation, reinterpretation, and self-correction of data in terms of theory, theory in terms of data."[46] No hermeneutic model of understanding is the appropriate one. It might be added that Thomas Kuhn's recent identification of paradigms (which grasp in global terms a particular way of apprehending the world) with the hermeneutic method reinforces this interpretation.[47]

In *Theorie des kommunikativen Handelns,* Habermas' account of scientism finds it proper place. The direction in which Habermas is heading has been apparent for some time. It was shown that in his debate with Niklas Luhmann, Habermas argues that it is neither science nor scientism, but functional thinking (which reduces every issue to that of system maintenance) that is the real villain.[48] *Theorie* develops this argument further. Habermas argues that under advanced capitalism the communicatively structured lifeworld is intruded upon by formally organized subsystems of purposive-rational action. The "unleashed functionalist reason of system maintenance disregards the claim of reason inherent in communicative socialization."[49] In a word, Max Weber's analysis of rationalization remains valid. Recall that for Weber rationalization signifies not merely that objectivities of all kinds have been demystified, so that there is no longer any source of transcendent meaning, but also that means have become ends in themselves. Science and bureaucracy, originally intended to make human action more effective in a complex and recalcitrant world, become ultimate values in themselves. Karl Löwith succinctly captures this process.

> This reversal marks the whole of modern civilisation, whose arrangements, institutions and activities are so 'rationalised' that whereas humanity once established itself within them, now it is they which enclose and determine humanity like an "iron cage." Human conduct, from which these institutions originally arose, must now in turn adapt to its own creation which has escaped the control of its creator.[50]

Or, as Habermas puts it, "the rationalized lifeworld enables the rise and growth of subsystems whose autonomous imperatives self-destructively strike back at it."[51]

Habermas' employment of Weber's analysis of modernity is not surprising. As was pointed out in chapter 1, Weber's concept of rationalization, especially as reinterpreted by Georg Lukács, is a central analytic principle of Horkheimer and Adorno's *Dialectic of Enlightenment.* However, unlike the first generation of critical theorists, Habermas is not utterly critical of the consequences of rationalization. The differentiation of institutional subsystems on the basis of different organizing principles (e.g., the organization of the economy along functional lines), characteristic of rationalization, has positive consequences. These include not merely a more dynamic economy but also at least the possibility of more fully developed individuals, as the relative autonomy of the modern family allows for a variety of modes of accommodation to the "reality principle." Indeed, much of *Theorie* is an assessment of the positive consequences of rationalization.[52] In *Theorie*, the problem of scientism falls into place. The imperialism of science is attributed not so much to its belief in itself as the only valid form of knowledge, but rather to the role it is called upon to play in the cultural drama of rationalization. In other words, Habermas no longer attributes to an epistemological doctrine—scientism—the power and influence of what he now regards as a social evolutionary force, rationalization. The result is that the validity of science is accepted within its own domain. The tendency of science to intrude upon communicatively structured aspects of the social world is attributed less to doctrines internal to the philosophy of science than to the exploitation of science by "expert cultures."

Habermas' hermeneutic understanding of the new philosophy of science, coupled with his recent interpretation of scientism as a social evolutionary (rather than primarily epistemological) force, would seem to render much of the criticism of his view of science in this and preceding chapters moot, i.e., out of date. However, Habermas' current position on the status of science and its philosophy is not entirely clear. In "A Reply," written apparently almost contemporaneously with *Theorie*, Habermas states that the technical cognitive interest continues to set the context of scientific knowledge.[53] One possible interpretation of this apparent discrepancy is that Habermas has tacitly split the philosophy of science and the practice of science almost completely in two under the influence of discourse. That is, the philosophy of science is seen by Habermas as a discursive, hermeneutic enterprise, but the practice of science remains bound to the technical cognitive interest. However, while this distinction seems to partially account for Habermas' position, it is not the whole story. One reason it is not is because it does not fit with Habermas' recent claim that ordinary science is free to interpret its data in a variety of ways, whereas reconstructive science is not. Were science as tightly bound to the technical cognitive interest as Apel points out it once was in Habermas' system (p. 91), the range of possible interpretations would seemingly be quite limited. At least these interpretations would all be of a single type. Perhaps what has happened is that the technical cognitive interest no longer has the force it once had in Habermas' system. This is presumably because Habermas no longer believes science itself need be contained by showing its epistemological roots in the species' struggle for survival. The technical cognitive interest rather serves, in Habermas' recent work, the more modest goal of standing as an epistemological barrier to a "new science" à la Marcuse.

The above interpretation is supported by the second reason why the critique of instrumental reason need no longer focus on science, according to Habermas. The claims to truth that science does appropriately make are to be taken seriously, he says. There is simply no alternative for those who would remain intellectually honest. In an interview on the occasion of the publication of *Theorie des kommunikativen Handelns*, Habermas states that when

> one devotes oneself to questions of truth and does not misunderstand himself in the process, then he should not, as Heidegger and Adorno have done, try to produce truths outside of the sciences or to set up a higher level of insight, on a memory of being or on a mindfulness of tormented nature. It is my deepest belief that one must not do that when engaging in thinking.[54]

For Habermas, as for Horkheimer, Adorno, and Marcuse, only reconciliation overcomes reification. However, for Habermas the only reconciliation possible is among men and women. It is sufficient if science understands itself modestly. It is then neither part of the problem nor part of the solution.

Yet, Habermas' position is—as always—complex. There has been in fact a slight appreciation of "nature romanticism" in his work for some time. This appreciation has taken the form of his somewhat more open attitude, in contrast to his position on Adorno and Marcuse, toward Walter Benjamin's idea of a mimetic, nonpurposively rational and spontaneous attitude toward nature, which is expressed in "post-auratic" art. Habermas' position is apparently that such an aesthetic attitude could perhaps exist alongside of an instrumental orientation.[55] In "Praktische Folgen des wissenschaftlich-technischen Fortschritts," Habermas states that the idea that human emancipation cannot take place without a resurrection of nature contains a kernel of truth; however, he does not elaborate.[56] Perhaps the idea of an artistic, aesthetic orientation toward nature, which would not take the place of but rather exist alongside of the technical orientation, expresses this kernel of truth for Habermas. For the most part, however, reconciliation is reserved in Habermas' system for those who can talk back. This is, he says, how Adorno's model of reconciliation should be reinterpreted. Otherwise reconciliation with nature requires too great a sacrifice of individual identity (via the mimetic identification with nature), a danger Adorno never quite escaped.[57] This issue is discussed in more detail in the next chapter. The point here is that while Habermas rejects the program of Horkheimer, Adorno, and Marcuse, there is a subtlety to his position that should not be overlooked. Indeed, perhaps it is even the case that Habermas' reconstructive science reflects his complex appreciation of the dialectic of Enlightenment. For reconstructive science, unlike ordinary science, claims not to impose its categories upon its objects, but to find its categories in them.

CHAPTER 9

Conclusion: Reconciliation with Nature or New Categories of Experience?

If one sees the core issue dividing Marcuse and Habermas as that of the possibility of a new science of an emancipated nature, then until quite recently few would have disagreed that Habermas had won the debate. However, in the last several years there has been considerably more questioning, if not dissatisfaction, with Habermas' solution to the dialectic of Enlightenment. The problem posed by the dialectic of Enlightenment, it will be recalled, is that scientific and technological progress, which create the material preconditions of a free society, require the repression of the erotic and playful aspects of human nature. Creating the objective material conditions of a free society destroys the subjective conditions of freedom. Two solutions to this dilemma are apparent. Marcuse hopes that at some point after the material conditions of a free society are present in abundance, socialist revolution could be coupled with a new orientation toward nature that would not require that man repress his own nature. This can be seen as a monistic solution, because the same reason that would foster fraternal relations among men and women would also foster a new fraternal relationship with nature. Habermas' solution, on the other hand, is dualistic, because he accepts the disenchantment of nature as the price of modernity, a price worth paying. Habermas divides reason in two, so that the emancipation of human relations need not require or depend upon the emancipation of nature.

Though few of Habermas' critics are impressed with Marcuse's solution, the dissatisfaction with Habermas' position, even among his most sympathetic critics, such as McCarthy, centers upon his dualistic solution.[1] McCarthy argues that Habermas' concept of a nature-in-itself plays an uncomfortable double role in his system. By the term "nature-in-itself" McCarthy apparently means that dimension of nature that exists independently of man's encounter with it as mediated by cognitive interests. The problem, according to McCarthy, is that nature-in-itself is not merely the object of science and technology. It is also the ground of human nature. However, McCarthy continues, if it is the case that we proceed from nature, how is it that our only access to nature is as though it were a completely foreign object—the object of the technical cognitive interest?[2] Is not Habermas' assumption that we know nature only via the technical cognitive interest really then a reflection of man's alienation from his own nature? (Habermas might reply that man's own nature is essentially—i.e., in those important respects that make him human—linguistic.) McCarthy sees in aspects of Habermas' earlier work, and particularly in his theory of social evolution, hints of "the possibility of a science of nature that would not be categorically rooted in an interest in technical controlability."[3] Hesse, referring to a passage in *Legitimation Crisis*, sees similar hints, this time of a science which would be more thoroughly integrated with meaning-giving world views.[4] It should be noted immediately, however, that Habermas rejects both of these interpretations of his work. While it is the case (as Habermas' response to Hesse in *Theorie des kommunikativen Handelns* suggests) that science has come to see itself in a hermeneutic fashion, Habermas states that for science to become part of the interpretative system of everyday life would require a regression behind modernity or the achievement of an entirely new cultural system. He does not consider the first desirable, nor the second likely: "The unity of reason has become unattainable at the level of cultural interpretative systems ever since the latter were detached in the modern period from the validity syndrome of everyday communication and split up under abstract individual aspects of validity."[5]

Habermas' rejection of McCarthy's interpretation is more complex, and can best be discussed by drawing upon the related criticisms of Henning Ottmann and Joel Whitebook.[6] McCarthy has reservations about Habermas' solution to the dialectic of Enlightenment because it makes permanent the fragmentation of reason. McCarthy believes that the "undifferentiated unity of earlier stages of thought [should be] reconstituted at the highest stages."[7] Habermas stops half way. However, if one asks why this reconstituted unity might be desirable (for reasons other than strictly philosophical ones having to do with contradictions in Habermas' materialism), then this question is best addressed by Ottmann and Whitebook. Ottmann equates the technical cognitive interest with the will-to-control nature.[8]* He goes on to argue that such an orientation leads to a revenge of external nature. As scientific and technological progress lead to environmental disruption, man seeks new scientific and technological solutions, which end up making the matter worse.[9] This problem, suggests Ottmann, will only be resolved by a new concept of nature as a "*purpose-for-itself*, in the face of which the will-to-control has to impose limitations on itself."[10] Unfortunately, says Ottmann, the technical cognitive interest prevents this. While Habermas recognizes the need for man to have a non-instrumental encounter with nature (e.g., an aesthetic attitude, as expressed in "post-auratic" art), he separates this orientation utterly from that of the technical cognitive interest.[11] What this means, implies Ottmann, is that there is no force in Habermas' system capable of binding to and limiting man's will-to-control nature. One important force that possibly could do this, an aesthetic orientation, is rendered impotent.[12] A certain parallel between Ottmann's criticism of Habermas, and Horkheimer and Adorno's critique of the dialectic of Enlightenment is apparent. In both cases, a force that could direct and control instrumental reason is elevated to higher realms and thus rendered powerless.

Habermas and Environmentalism

It is interesting that some of the sharpest criticisms of Habermas' scheme, at least among sympathetic critics, stem from its perceived

*Though Habermas would not agree with this equation per se, it does point to the fact that Adorno's interpretation of modern man's relationship to nature as one of "wild self-assertion" [*verwilderte Selbstbehauptung*] remains an influence upon Habermas' scheme. Indeed, one can see the tech-

inability to fully address today's environmental problems. Indeed, one might regard these environmental problems as the contemporary version of the rise of totalitarianism in the 1920s and 1930s, insofar as (and only insofar as) the persistence of these problems may also point to a flaw in the Enlightenment concept of reason. For as Whitebook points out, Habermas seeks the successful completion of the Enlightenment project. Habermas "takes rational autonomy as a perfectly adequate idea of selfhood, and feels no need to formulate an alternative emancipatory concept such as a new sensibility."[13] However, if today's environmental problems should prove intractable, then it could readily be hypothesized that there is something wrong with the Enlightenment project itself.[14] Man's domineering relationship to nature would testify to the inadequacy of the Enlightenment concept of reason in the latter half of the twentieth century, as man's inhumanity to man (itself hardly overcome!) testified to its inadequacy in the first half of the twentieth century.

Whitebook is cautious in formulating his treatment of this issue. Indeed, he formulates Habermas' reply so well that Habermas endorses it.[15] Whitebook points out that communicative ethics is thoroughly anthropocentric. By virtue of his communicative abilities, man is the only value-bearing being. This in itself is sufficient to rule out any conception of nature as an "end in itself."[16] Habermas' claim, under the rubric of the technical cognitive interest, that nature in itself is unknowable simply reinforces this position. Habermas' system secures dignity for men and women by virtue of their potential to communicate. However, the way in which Habermas secures human dignity requires that he deny all worth to nature. The "good-for-nature" must be solely derived from the "good-for-man" in Habermas' system.[17] The question, says Whitebook—and he does indeed intend it as a question—is whether we can "*continue to deny all worth to nature and treat it as a mere means without destroying the natural preconditions for the existence of subjects*? Likewise, can the [worth] of nature be secured without devaluing the dignity of the subject?"[18]

Whitebook goes on to review how Habermas might argue that

---

nical cognitive interest as not merely neutralizing this orientation (i.e., making it a species interest, and thus more abstract), but as legitimating it by confining it to nature. Indeed, this is the point of Habermas' dualistic strategy.

the "good-for-nature" is somehow entailed by the communicatively conceived "good-for-man." (This argument will be addressed shortly.) Whitebook concludes that even if it could be formally demonstrated that a discourse ethic could generate "binding imperatives which oblige us to protect species from extinction," and the like, this might well be practically irrelevant.[19] The key issue may well be one of social psychology. A change in the modern world view—which sees nature merely as the field of man's endeavors—may be necessary to mobilize public opinion sufficiently to redirect scientific and technological progress away from an ultimate collision with nature. Interestingly, in his theory of social evolution Habermas has told us how this type of leap to a higher level of learning, which would draw upon the resources contained in the world views of previously marginal groups, might occur. However, suggests Whitebook, Habermas' system, which denies that alternative attitudes toward nature could ever have theoretical efficacy, seems to work against this possibility.

In a moment Habermas' response will be noted. However, it should not be overlooked that while the criticisms of McCarthy, Hesse, Ottmann, and Whitebook are suggestive rather than definitive, and not always precisely formulated, they focus upon a common theme. Habermas seems to build too many divisions into his system. In particular, the quasi-transcendental character of his scheme continues to deny that new knowledge and experience could alter the dualism of his fundamental categories. Originally this was done, I suggested, to protect the autonomy of critical theory, and particularly to protect its claim to be the special science of human development in the realm of language and morals. However, it may be that the project of securely grounding critical theory, once so theoretically fruitful, is beginning to exact too great a price. If this should be the case, then what is really the greatest defect of a quasi-transcendental scheme such as Habermas' becomes apparent. His scheme should be a source of concern not because it cannot guarantee knowledge—Habermas was never under the illusion that it could—but because it could act as an impediment to the growth of new knowledge, understood as radically new categories of experience.

Yet, Habermas' response to McCarthy, Ottmann, and Whitebook is trenchant. In particular, he argues that the limits of a discourse ethic vis-à-vis nature are not so entirely different from the limits of this ethic vis-à-vis other generations. Critical theory, it will be recalled, has always made a special effort to come to terms with human suffering. Adorno suggests that there could be no poetry after Auschwitz, and Marcuse at one point claims that the aesthetic form could redeem human suffering, but he later argues that the aesthetic form could only memorialize it.[20] Similarly, Habermas claims that there remains a "stain on the idea of justice that is bought with the irrevocable injustice perpetrated on earlier generations. . . . The contradiction that is inherent in the idea of complete justice, owing to its in principle irredeemable universalism, cannot be dissolved."[21] Indeed, the ethical difficulty with a discourse ethic is not merely that those who have suffered and died cannot share in the solidarity of consensus. It is also that the participants in discourse must make decisions that will profoundly affect those who cannot participate in discourse. Generations yet to come are a prime example. Yet, even though the unborn cannot even counterfactually be said to participate in discourse, there is no reason to presume that they will be regarded as mere objects. Similarly perhaps with nature.[22] Habermas' point is simple and powerful: Ottmann and others confound epistemological and ethical issues.[23] Because the only "theoretically fruitful attitude" toward nature is one bounded by the technical cognitive interest means neither that other noncognitive (e.g., aesthetic) attitudes are impossible nor that man must run roughshod over nature.

To be sure, a recent version of what might be regarded as a discourse ethic has run into difficulty in this regard. Bruce Ackerman, in *Social Justice in the Liberal State*, proposes what may be called neutral dialogue. It is intended, not unlike Habermas' scheme, to realize the radical potential of the liberal democratic ideal of free speech. Ackerman's idea, to put it very simply, is that any distribution of resources is to be considered justified unless it can be criticized on grounds that do *not* recur to the superiority of the criticizing individual or his ideas. However, this scheme leads to results that are ethically counterintuitive to many. In particular, those who cannot participate in neutral dialogue, those who cannot raise neutral objections, have no rights. Thus, in Ackerman's example, there is little to be said against someone who took pleasure from infanticide, unless doing so happened to violate the "adoption rights" of other participants in dialogue.[24]*

*Ackerman claims that he has established arguments against wanton cruelty in chap. 3, sec. 22.2. But, as William Galston points out in a review in *Political Theory*, the prohibition is not mandatory, but optional.

Habermas' scheme, precisely because it lays claim to a much more ambitious concept of speech as intending truth, avoids this difficulty, at least in theory. The telos of truth implicit in every speech act, which is allowed to emerge when the formal conditions of discourse are met, suggests that agreement in discourse is more than mere agreement among equals. Such agreement lays claim to the status of an agreement that would be universally assented to by all those whom it would affect, now or in the future.[25] Discourse claims to be conducted in the name of all mankind. Though it cannot ever claim to have adequately represented all who will be affected, it seeks universal assent. This understanding of discourse captures quite well the emancipatory ideal of the first generation of critical theorists: to speak for and in the name of those who cannot speak for themselves. In fact, Habermas emphasizes the embeddedness of discourse in this tradition.[26]

Within such a framework, Habermas implies that the sufferings of animals as well as humans can hardly be ignored. To deny animals epistemological priority with humans is not to deny them ethical significance.

> The impulse to provide assistance to wounded and debased creatures, to have solidarity with them, the compassion for their torments, abhorrence of the naked instrumentalisation of nature for the purposes which are ours but not its, in short the intuitions which ethics of compassion place with undeniable right in the foreground, cannot be anthropocentrically blended out.[27]

Though "anthropocentrically blended out" is hardly the most felicitous of expressions (or translations), the idea is clear. Though formal theoretical questions regarding such things as justice for animals perhaps place too great a burden on a discourse ethic, compassion and decency do not. Discourse explicates these ethical intuitions. It does not grant them their power or confine them to those who can participate in their explication. In "Psychic Thermidor and the Rebirth of Rebellious Subjectivity," Habermas states that in his last philosophical discussion with Marcuse, shortly be-

---

Thus, if a liberal society does not feel like restraining wanton cruelty, infanticide—even of normal children whose parents can afford to rear them—is ok, except in the special circumstances in which the adoption rights of other, infertile adults, are thereby violated (p. 432).

fore Marcuse's death, Marcuse stated that he knew where our most basic value judgments are rooted: "in compassion, in our sense for the suffering of others."[28] Whether or not this is also Habermas' view, it is entirely compatible with his discourse ethic. Discourse would help explain and justify such values. It is neither their source nor their limit.

If treating nature with respect and dignity does not require that it be regarded as an end in itself, if Habermas' scheme need not lead to the revenge of external nature, is there any reason to remain dissatisfied with his scheme? One could respond that even if Habermas' scheme need not lead to environmental disruption, we nonetheless require a new myth of man's unity with nature in order to mobilize the opinions and emotions of the masses. A justification of environmental sensitivity on the basis of a discourse ethic would be reserved for intellectuals. However, such a strategy would seem to require an unacceptable degree of manipulation, or pandering. In "The Classical Doctrine of Politics," Habermas argues that for centuries religion served as the philosophy of the masses.[29] However, religion and philosophy shared common assumptions. Indeed, it could be argued that until the modern age one was the simplification of the other in many respects. A myth of man's unity with nature, designed for popular consumption and fostered for social psychological reasons, would be a far more cynical undertaking.

One could also argue, in response to the question of why Habermas' scheme remains unsatisfying, that it does not address the revenge of *internal* nature; i.e., the malaise that results from the repression of those aspects of human nature that seek peace and joy. This is a much more difficult question, an aspect of which will be addressed later. For now it must suffice to note that if Marcuse is correct about the sacrifices of human nature required to dominate world nature, then McCarthy's position that Habermas' dualistic scheme is best regarded as a way station on the road to a higher unity would seem to be correct. Unless, that is, it could be shown that the only alternative was a regressive unity; i.e., there was nothing acceptable to be done about this sacrifice.

The possibility of a higher unity of man and nature, somewhat along the lines of Marcuse's speculation, cannot be pursued further here. However, if I am correct, a certain dissatisfaction with Habermas' scheme expressed by McCarthy, Ottmann, and Whitebook can fruitfully be seen from a slightly different perspective. From this perspective, the difficulty with Habermas' scheme is not that it does not speculate about what has not yet been achieved and may well never be: a higher unity. The difficulty is rather a certain fixity or rigidity in Habermas' categories, which mitigates against speculation. In particular, Habermas' system downplays the creative freedom with which man constructs his philosophy and science. Habermas' system seeks in some measure to fix in advance the categories of what man does make: the intellectual artifice of his culture. Reinhart Maurer was one of the first to focus on this point. As was noted previously (p. 92), Maurer states that even after the introduction of discourse there is no theoretical distance in Habermas' scheme between self-reflection and interests. Interests determine self-reflection. Discourse is simply the "mediation" of interests.[30] While Maurer exaggerates the rigidity of Habermas' position (it is through academic discourse that Habermas has been led to abandon aspects of cognitive interest theory), it is the case that Habermas takes pains to prevent new knowledge from feeding back to modify the status of interests or, later, the special claims of reconstructive science. The way in which this contributes to a certain inflexibility in Habermas' system is considered below. This issue will be approached by considering in more detail Habermas' claim that toward nature there exists "only one *theoretically fruitful* attitude, namely the objectivating attitude of the natural-scientific experimenting observer."[31]

Before considering this certain inflexibility in Habermas' system, however, it is important to be clear about what Habermas is and is not claiming. Russell Keat argues that Habermas is "resolutely opposed to a 'hermeneutics of nature,' to a science of nature that is interpretative rather than empirical."[32] Keat sees this as inconsistent with Habermas' doctrine of cognitive interests. If cognitive interests are constitutive of reality, determining the aspects under which reality is objectified and made accessible to experience to begin with, then a hermeneutics of nature should be quite possible. That is, if the object domains of the practical and technical interests are defined almost entirely in terms of the interests themselves, then why could the interests not cross over, such that one would also have a science of the practical and a hermeneutics of nature? If interests are reality constitutive, how could nature object or resist? Keat concludes that Habermas is inconsistent: the impossibility of a hermeneutics of nature invalidates the reality constitutive character of cognitive interests.[33]

Keat, however, does not formulate the issue with sufficient precision. Habermas does not deny the possibility of a hermeneutics of nature at all. In fact, Habermas is somewhat attracted to a version of this idea, especially as expressed by Walter Benjamin, as has been shown. What Habermas rejects is rather that noninstrumental attitudes toward nature could ever have "theoretical efficacy." By this term Habermas seems to mean that such attitudes toward nature are incapable of acting on nature in such a way as to meet humankind's physical needs. It is this claim of Habermas' that is examined below. Habermas may be mistaken, but he is not guilty of the internal inconsistency with which he is charged by Keat. Furthermore, our earlier discussion of cognitive interest theory suggests that Keat interprets interests far too idealistically. Habermas believes that the objects of interests constrain interests. Habermas also believes that human nature constrains interests.[34] Interests may be reality constitutive, but as Reinhart Maurer points out, we appear to have little choice as to how they constitute reality.[35]

Four reasons on behalf of Habermas' claim that toward nature there exists "only one *theoretically fruitful* attitude," namely that of the objective scientist, can be found in his recent work. The fourth reason is the most fundamental, and to some degree it summarizes the other three. All four reasons supplement and modify but do not abandon Habermas' division of the world, under the rubric cognitive interest theory, into the realms of labor (the technical cognitive interest) and language (the practical cognitive interest).

1. Habermas believes that the disenchantment of the world characteristic of the Enlightenment represents a process that is structurally similar to that of individual cognitive development, which is characterized by a gradual "decentering of an egocentrically expressed world understanding," as described by Jean Piaget. The "desocialization of nature," coupled with the "denaturalization of the human world," is thus a progressive movement corresponding to individual maturation, in which the world is no longer seen simply as an extension of human hopes and needs.[36] Problems associated with this position were examined in the last chapter.

2. Habermas believes that the general understanding of nature held by modern science is accurate, or at least that there exist no "higher" truths about nature that might be grasped by intuitive methods. It is in this context that Habermas notes that Adorno's concept of the remembrance of nature comes "shockingly close" [*schockierende Nähe*] to Heidegger's concept of the remembrance of being. Both abandon the goal of cognitive, theoretical knowledge.[37]

3. Habermas understands modernity in terms of the increasing differentiation of three moments of reason [*Vernunftmomente*]: modern science and technology; positive law and secular ethics; autonomous art. In so doing, he states that he is following Max Weber's analysis, in which modernity is characterized by the breakup (disenchantment) of a religiously expressed world view into separate and autonomous spheres of value.[38] By defining modernity in terms of the increasing differentiation of world views, Habermas renders—virtually by definition—the attempt to achieve a reintegration of these "moments of reason" (e.g., in a new science along the lines discussed by Marcuse) a regressive act.

4. Habermas believes that this "*differentiation of cultural spheres of value is due to special family relationships between basic formal-pragmatic attitudes* (that are rendered methodologically useful) and *corresponding domains of reality*."[39] If I am not mistaken (Habermas' claim is not utterly transparent), Habermas is simply stating that some ways of knowing are more effective when applied to some domains of reality than others; e.g., literary criticism is better suited to enhancing our knowledge of a text than our knowledge of nature. Though it is perhaps the case, suggests Habermas, that all ways of knowing are applicable to all domains of reality at some level, such as the scientific study of art, ways of knowing are much more efficacious when applied to the appropriate region of reality. (It is this distinction that Keat misses.) This seems to be due to a certain structural similarity, or fit, between ways of knowing and region of reality. Habermas develops this claim further in his statement that "the moralisation of our dealings with external nature cannot, any more than can a non-objectivating knowledge of nature, be carried out *at the same level* that Kant attained in his moralisation of social relations and Newton attained in his objectivating knowledge of nature."[40]

It should not be overlooked that reason four is in fact an assertion. (I focus on this reason, as it seems the most fundamental; reason one has been thoroughly criticized by McCarthy and Schmid, while reasons two and three seem to be more akin to position statements, whose validity depends in significant measure on reason four.) It is a systematic assertion—i.e., it organizes several claims into a pattern—but it remains an assertion. It is, for example, hard to imagine what sort of evidence could be appealed to (especially now that Habermas has explicitly abandoned the strategy of transcendental deduction)[41] in order to demonstrate that the current relationship between ways of knowing and regions of reality is not merely contingent. Now, it seems quite likely that all one can do in this realm is to make assertions. Arguments in favor of a particular way of organizing reality, at least at the fundamental level Habermas is concerned with, are perhaps not so much true or false as they are fruitful or nonfruitful in calling attention to new and significant aspects of things. Or, as Nelson Goodman puts it: "For a categorical system, what needs to be shown is not that it is true but what it can do. Put crassly, what is called for in such cases is less like arguing than selling."[42] What should be considered is perhaps not so much the evidence (which must ultimately be meager) that Habermas brings to bear in favor of his claim regarding the only "theoretically fruitful attitude," but whether this claim—and the scheme it is based on—helps us see the world in new, more fruitful and interesting ways; i.e., does it sell? Or does it constrain the imagination unnecessarily?

The first thing to notice in this regard is that Habermas' claim seems to be an artifact of the way in which he employs diagrams to represent his argument. One sees this especially in "Schemas 1 and 2" in "A Reply to my Critics," but it is also apparent in similar diagrams in *Theorie*.[43] In these diagrams, Habermas places various attitudes or orientations toward reality on one axis and the various regions of reality on the other. Then he fills in the cells. "Schema 1" in "A Reply" (p. 244) is exemplary.

FORMAL-PRAGMATIC RELATIONS

Basic Attitudes	*Domains of Reality*		
	(1) External nature	*(2) Society*	*(3) Internal nature*
(1) Objectivating	*cognitive-instrumental*	*cognitive-strategic*	*self-objectification*
(2) Norm-Conformative	*moral-aesthetic relation to non-objectivated environment*	*obligatory relation*	*self-censoring*
(3) Expressive	*same as above*	*self-presentation*	*sensuous-spontaneous relation to self*

Is it surprising that such a way of representing the relationship between ways of knowing and domains of reality leads to very little combining or blending of "basic attitudes"? If a particular cell is difficult to fill in, as a "norm conformative" attitude toward nature surely must be, then it becomes a virtual conceptual nonentity.

> The *norm-conformative attitude* to this domain of external nature does not yield any problems susceptible to being worked up cognitively, that is, problems that could be stylised to questions of justice from the standpoint of normative validity.[44]

Is this conclusion not simply a product of the way in which Habermas conceptualizes the issue? One could respond, correctly, that all conclusions are generated by their premises. However, the point here is substantive. Not only does Habermas' particular way of conceptualizing the issue make any blending of "basic attitudes" virtually impossible, but it also tends to render nonobjectivating attitudes toward nature somewhat silly. This is because such attitudes are derived from attitudes designed to explain human relations and applied to nature merely by virtue of the symmetrical structure of Habermas' conceptual scheme. One might also respond that it is misleading to focus upon an author's diagrams (though it is surely the case that Habermas is fond of diagrams, as thumbing through *Theorie* will readily show). Diagrams often render distinctions more sharply than their literary discussion. There would be some truth in such an objection. Earlier, in "A Postscript to *Knowledge and Human Interests*," Habermas pointed out that in practice instrumental and communicative action are always blended, e.g., science is a social enterprise in which scientists must talk with each other, as well as manipulate nature.[45] Nevertheless, it can hardly be denied that Habermas intends to emphasize the fundamental duality of attitudes and that he does so for a purpose. Earlier this purpose was to combat a reductive Marxism as well as an agressive positivism. Today this duality also serves to combat the monism of the dialectic of Enlightenment.

Just as Habermas states that in practice instrumental and communicative action are almost always blended, he also states that in everyday life a certain "unity of reason" already occurs. "The unity of reason that McCarthy, Ottmann and others invoke is not to be had at the level of the cultural spheres of value. On the other hand, in the communicative practice of everyday life, in which cognitive explanations, moral expectations, expressions and evaluations interpenetrate, this unity is in a certain way *always already* established."[46] Habermas does not elaborate. He seems to mean that while the different attitudes toward reality cannot be integrated without a loss of efficacy, it is nonetheless the case that in everyday life, when a decision is made to build a dam, a bomb, or clean up the air, all the different attitudes will come into play. However, while this position is certainly reasonable, it is perhaps also misleading, insofar as it suggests that "cultural spheres of value" live an autonomous existence in the first place. That is, Habermas allows for no feedback to occur from social practice to basic attitudes.[47] The differentiation of attitudes is given (except at the cost of regression). Attitudes are simply mediated more or less adequately by social practice. However, if a categorical scheme is to be judged in terms of its fruitfulness and the like, what purpose is served by a scheme that excludes in advance the possibility that new practices might constitute new attitudes, or new combinations of attitudes? This point has been made in a somewhat different context (referring primarily to new social movements, rather than to basic attitudes) by Seyla Benhabib, and Murray Bookchin.[48] Benhabib puts it this way.

> It is not clear why a new socialization of the individual beyond the patriarchal family, school, and culture, and a new mode of material interaction with nature beyond the industrial mode of production, would be impossible. No theory can define the limits of future possibility, although it can enlighten us about it. For this possibility is posterior and not prior to actuality—as Aristotle long ago said of praxis.[49]

Blended Basic Attitudes

Until this point the argument has focused on the fact that Habermas seeks to exclude in advance a range of abstract possibilities regarding the relationship between humanity and nature. It is now appropriate to consider some of these possibilities in more detail. The first thing to notice in this regard is that it is by no means obvious that it is even currently the case that nonobjectivating attitudes toward nature lack theoretical fruitfulness. The field of medical anthropology is replete with case studies of noninstrumental healing (or harming) techniques that work. The much studied phenomenon of voodoo death is perhaps the most dramatic example. The diagnostic efficacy of Navaho hand tremblers is also frequently cited.[50]

It might be argued in response that the fact that these techniques frequently work is quite a different issue from that of their theoretical efficacy. Though Habermas never defines what constitutes "theoretically fruitful knowledge," presumably it denotes more than merely techniques that work. Voodoo, after all, works (at least if one believes in it). "Theoretically fruitful knowledge" apparently also means knowledge in which particular healing techniques, for example, are part of a larger explanatory structure from which these techniques can be predicted, derived, and explained. However, even a passing familiarity with voodoo or Navaho healing techniques shows that these techniques are hardly isolated tricks. Voodoo and Navaho healing techniques are part of a sophisticated explanatory structure that explains why they work, as well as the limits of their applicability. This has been shown by the studies of Lévi-Strauss and William Morgan, among others.[51] Indeed, such techniques would hardly have captured the interest of countless anthropologists were they not part of a larger world view.

It could be replied that this larger world view is not a scientific theory, because it frequently refers to religious and mythological entities, lacks the requisite precision, and so forth. Hence, these techniques can never be theoretically fruitful. However, such a reply would be no more than an attempt to deal with this complicated issue by definitional fiat. Such a reply would be tantamount to the assertion that nonscientific techniques are not scientific because they do not rest upon scientific theories. Such an assertion is true but trivial. The fact that "primitive" techniques are rooted in a complex structure of thought should be sufficient to establish their potential theoretical fruitfulness, if this term is to have any meaning. Actually, what Habermas seems to mean by the claim that only the scientific approach to nature is "theoretically fruitful" is more general, but no less problematic. He seems to mean that only the scientific approach has cognitive and intellectual content. Other approaches are feeling-based, aesthetically oriented, or emotional.[52] That "primitive" healing techniques also serve as a counterexample against this more general claim is apparent.

In response to the preceding, one might argue the obverse: that "primitive" healing techniques work, when they do, because they are actually based upon scientific medicine. Though the witch doctor (to indulge in a stereotype for a moment) may attribute magical properties to the bark of a particular tree, we know that it possesses medicinal properties because it contains quinine. The claim that an instrumental relationship to nature is the only theoretically fruitful one is apparently upheld by "translating" the traditional practice into objectivistic terms. However, much faith healing, including that practiced by Western psychiatry, is not so readily and completely subsumed by objectivistic thinking. Stomach ulcers can indeed, it appears, be cured by taking a "communicative attitude" toward nature; i.e., psychoanalytic talk therapy. As with the case of voodoo death, the efficacy of talk therapy can be explained in objectivistic terms, in both cases in terms of the effect of stress upon the nervous system. But it cannot be explained away. It remains the case that an aspect of nature may be profoundly altered by a communicative attitude aimed at "consciousness" and "convictions," terms Habermas once used to explain the efficacy of rhetoric.[53] It might be objected that human physical nature is a special case. The mind and its body are in uniquely intimate contact, quite unlike the relationship between mind and external nature. However, it is McCarthy's point, after all, that precisely because men and women stem from nature and thus are part of nature, that the alienation from nature implicit in Habermas' dualistic categories is so problematic. Furthermore, according to the critique of the dialectic of Enlightenment, it is man's self-imposed alienation from his own nature that is the greatest cost of progress. In this light perhaps the efficacy of noninstrumental healing (and harming) methods is not an exception, but the heart of the matter.

It has, of course, all along been Habermas' position that self-reflection, based upon the psychoanalytic talk therapy model, is the paradigm of how human emancipation may be won.[54] However, all along Habermas has separated the psychoanalytic paradigm from the issues considered here; i.e., mind-body interaction. To be sure, Habermas recognizes that psychoanalytic self-reflection is in large measure about the rationalization and integration of needs and drives, which presumably stem in large measure from man's nature. However, Habermas focuses almost exclusively upon the linguistic expression and cognitive understanding of needs and drives. As Ottmann points out, Habermas over-intellectualizes the process of psychoanalytic reflection.

> It seems exaggerated to elevate the patient's "self-reflection" to a means of liberation. In psychotherapy, liberation is more

> the result of the "emotional acting-out of the conflict," of repetition, resistance, and emotional upset. . . . In Habermas' intellectualised interpretation, reflection is attributed to what is actually accomplished by the working out of the conflict.[55]

In a related fashion, Russell Keat argues that Habermas is quite mistaken in his interpretation of psychoanalysis to the effect that "Id = alienated ego."

> Having (mis)-understood the concept of the id as the alienated ego, he [Habermas] presents in effect a literal and unqualified reading of this dictum ["Where Id was there Ego shall be"], so that the abolition of the id is seen as a possible and desirable outcome of the therapeutic process. Likewise, the instincts are regarded as the sources only of pathological neurotic activity; and indeed the same is true of all unconscious determinants.[56]

Though Keat exaggerates, there does seem to be a certain alienation from aspects of human nature implicit in Habermas' concept of psychoanalysis, insofar as there seems to be little space for those aspects of human nature that cannot be made utterly transparent in language. The conclusion to be drawn is, I believe, that while there is certainly a place for psychoanalysis in Habermas' system, there is no adequate place—no appropriate category—for the type of "primitive" phenomena discussed above, phenomena that seem to be most similar to a communicative attitude toward nature, but do not quite fit this category either. Taking the manifold complexity and diversity of the world seriously seems to overload Habermas' categories.

Finally, it might be replied that the need to have recourse to the anthropological literature makes Habermas' point. One must, it seems, recur to examples from primitive tribes and the like to find examples of blended "basic attitudes" (as defined in terms of Habermas' scheme on p. 150). This would inadvertently demonstrate Habermas' point that there can be no blending without regression. However, the story regarding these "primitive" tribes is not so simple.[57] Anthropologists sometimes speak of the "150 percent man." The idea is that members of a traditional society, when confronted with the incursion of modern society (acculturation) do not simply abandon old beliefs for new ones, as in the emptying and refilling of a vessel. Rather, they add new beliefs to the old, mixing and matching to suit their needs. Quite arguably this combination is richer than either the traditional or modern culture alone. The field of medical anthropology again provides examples. In general, the pattern in many traditional societies seems to be that mental illness and chronic physical illness are believed to be best treated by traditional methods, whereas acute illness is often thought best treated by modern methods. The function of Navaho hand tremblers as mediators between traditional and modern methods is interesting in this regard. "Navaho curers respect White doctors, acknowledge their superiority in some areas, and cannot understand why their respect is not reciprocated. Diagnosticians (hand tremblers) are still usually the first professionals to be consulted and they in turn advise 'their patients to go to the clinic in certain circumstances, rather than to the medicine man.'"[58] Nor is it the case that traditional societies, having experienced firsthand the efficacy of modern scientific medicine, progressively abandon the old ways. "Counter-acculturation" is not uncommon. However, as Lola Romanucci-Ross has shown in her study of the Manus culture (of the Admirality Islands), counter-acculturation does not necessarily involve the total rejection of modern methods either. It can be seen as an attempt to bring modern and traditional methods into a more satisfactory balance.[59]

Habermas might respond that if a blending of healing approaches were desired in modern societies, perhaps because such a blending seems to strike a nice balance between instrumental and communicative attitudes toward nature, then this blending could occur at the level of the "communicative practice of everyday life." However, while the dualism of basic attitudes can be salvaged in such a fashion (What apparent evidence of the blending of basic attitudes in modern society could not, after all, be reinterpreted this way?), what does it matter if the blending of healing approaches in the Manus culture or our own takes place at the level of communicative practice or of world views ("cultural spheres of value")? Can this distinction really be upheld? If so, Habermas develops no argument in its defense. One is forced to ask whether the primary function of this distinction is not to preserve Habermas' basic categories. Yet, this distinction certainly serves a purpose within the framework of Habermas' project. The denial of a possible blending of basic attitudes—the point of Habermas' distinction between practice and world view—is central to Habermas' interpretation of the dialectic of Enlightenment in terms of a discourse ethic. Only such a reinterpretation, Habermas seems to believe, can save critical theory from irrelevance.

Reason and Reconciliation

Without reviewing the critique of the dialectic of Enlightenment further, it can be noted that Seyla Benhabib seems correct that the *Dialectic of Enlightenment* states the problem of Western thought in such profound terms that conceptual thought itself becomes the problem.[60] Western thought, according to Horkheimer and Adorno, exhibits "identity logic," in which things go into their concepts without remainder. Reason seeks to control via abstraction and generalization. Abstraction and generalization in turn grasp reality only by liquidating the unique individuality, the particularity of the other. However, if this is so—if conceptual thought is itself the problem—then what is the alternative? For Horkheimer and Adorno, art and the doctrine of mimesis seem to be the answer. Benhabib discusses the way in which art transcends "identity thinking" quite clearly. As for mimesis, Habermas argues that it is "the pure opposite" [*das bare Gegenteil*] of reason; it is pure impulse. Mimesis, says Habermas, implies a snuggling [*anschmiegen*], imitative, highly sympathetic relationship, one which when applied to nature is little more than a "cipher" [*Chiffre*].[61] That is, it is very unclear what sort of actual relationship to nature could give mimesis content, unless one thinks in terms of hugging one's housepet and the like. For Habermas it is equally unclear how a mimetic orientation could be blended with a cognitive, theoretical one, insofar as theoretical knowledge is itself the problem for the dialectic of Enlightenment.

> As the placeholder for this primordial reason that was diverted from the intention of truth, Horkheimer and Adorno nominate a capacity, *mimesis*, about which they can speak only as they would about a piece of uncomprehended nature. They characterize the mimetic capacity, in which an instrumentalized nature makes its speechless accusations, as an "impulse." The paradox in which the critique of instrumental reason is entangled, and which stubbornly resists even the most supple dialectic, consists then in this: Horkheimer and

> Adorno would have to put forward a *theory* of mimesis, which, according to their own ideas, is impossible.[62]

Before considering further Habermas' attempt to reinterpret mimesis, it may be useful for a moment to consider why Habermas would see mimesis as utterly opposed to theoretical reason in the first place. Habermas appears to ignore what Adorno seems to have regarded as a virtual second dimension of mimesis: its active, constructive moment. One sees this second dimension of mimesis most clearly in what Adorno calls "exact fantasy."

> An exact fantasy; fantasy which abides strictly within the material which the sciences present to it, and reaches beyond them only in the smallest aspects of their arrangement: aspects, granted, which fantasy itself must originally generate.[63]

As Susan Buck-Morss points out, exact fantasy is mimetic in that it lets the object (the facts presented by science in this case) take the lead, set the stage. While the subject's imagination intervenes to create something new, it is at the same time guided and constrained by the object. Literary translation and musical performance are similarly mimetic. They do not merely copy the original. They maintain the "aura"—the presence—of the original by transforming it in the very process of reproduction.

For Adorno, mimesis has very little to do with a direct and unmediated encounter with nature. Such a direct encounter, were it even possible, would be tantamount to the fetishization of nature. Abstracted from the whole, which includes its social context, the natural object "congeals . . . into a fetish which merely encloses itself all the more deeply within its existence."[64] In fact, our experience of nature is always mediated by history, culture, and science. The primacy of the particular that mimesis connotes does not refer to the primacy of the object per se. It rather refers to Adorno's attempt to substitute a constellation of mediating factors for the intellectually lazy practice of apprehending the object by simply subsuming it as an instance of familiar categories. Perhaps the most dramatic way in which Adorno sought to abolish conceptual hierarchy is in his own paratactic literary style, which places elements in opposition rather than arguing from the general to the particular or vice-versa.

Why would Habermas, who knows all this and more, downplay

the complexity of mimesis, especially its active and constructive character? Since Habermas cannot be unaware of this aspect of mimesis, the answer must be that the active, constructive dimension of mimesis is not seen by Habermas as being theoretically significant. Why this might be the case is found in Habermas' claim that Adorno abandoned theory altogether. He "practiced ad hoc determinate negation."[65] What this seems to mean for Habermas is that Adorno never brings the active and receptive, the constructive and imitative, together in the form of a higher synthesis. Rather, mimesis emphasizes one element, then another, in a sort of force-field characterized by an alternating current illuminating first the active, then the receptive dimension of experience, but never both at once.

Why such a concept of mimesis would be for Habermas the "pure opposite" of reason—i.e., pure impulse—is apparent. For Habermas, reason guides, synthesizes, and categorizes. To abandon this dimension of reason, even half-time as it were, is tantamount to the abandonment of reason itself. A concept of reason that even temporarily lets itself be guided by the object is hardly reason at all for Habermas. To be sure, Habermas does not put it quite this way. Rather, he writes of Adorno as perpetually residing within the "paradoxes of a self-negating philosophy," which a scholar—as opposed to Adorno the literary artist—cannot long live with.[66] It appears that it is the power of Habermas' own system, his own understanding of reason as categorizing and synthetic, that leads him to see as relatively insignificant the active, conceptualizing, and constructive dimension of mimesis, precisely because these elements never take the lead, at least not permanently. Just as "primitive" healing techniques are not "theoretically fruitful knowledge" by virtue of how Habermas defines the latter term, so mimesis is the antithesis of reason by virtue of how Habermas defines reason. Though the definition of reason at issue here is tacit rather than explicit, the idea is clear: reason does not let itself be ordered and directed by experience; it takes the lead. One can readily trace this understanding of reason back to cognitive interest theory, in which the transcendental elements (interests) predominate, even as the objective world sets outer limits. At first glance this arrangement might seem similar to Adorno's "exact fantasy," but upon reflection the hierarchical relationship between interests and objectivity is quite apparent.

A central intellectual move in *Theorie* is Habermas' reinterpretation of the mimetic relationship in terms of his communications theory. Drawing upon his discourse ethic, Habermas reinterprets mimesis as pure sympathetic identification with another person, culminating in an understanding between them utterly free of compulsion. Habermas quotes from Adorno's interpretation of Eichendorff's concept of "beautiful otherness" [*Schönen Fremde*], in order to capture the concept of reconciliation as applied to human relations.

> The situation of reconciliation does not annex the foreign as a form of philosophical imperialism; its happiness stems from its protected nearness to the distant and different, on the other side of the heterogeneous as well as individual.[67]

Though he did not intend to, says Habermas, Adorno actually described reconciliation in terms of unimpaired intersubjectivity. That is, Adorno in effect describes reconciliation in terms of an intersubjectivity that is established and maintained in the reciprocity of mutual understanding based upon free recognition. Mimesis lends itself beautifully to—indeed, it seems to invite—reinterpretation strictly in terms of intersubjective communication between people. In this reinterpretation nature no longer has any role to play. The gains in rigor, clarity, and systematic power that result from this move are dramatic. At the same time there are losses, subtle ones to be sure, that should not be overlooked.

Habermas' reworking of the *Dialectic of Enlightenment* in terms of a communications philosophy (which he describes in terms of a "paradigm change" from a consciousness philosophy to a communications philosophy) is a brilliant, and in large measure successful, attempt to reestablish the interdisciplinary character of critical theory, giving it renewed theoretical focus. Yet, our considerations suggest that Habermas goes too far, or at least that he unnecessarily forecloses a number of options. Even should one agree with Habermas that the *Dialectic of Enlightenment* is led into a cul-de-sac by its insistence on reconciliation with nature, this need not imply that the acceptance of a strictly instrumental relationship with nature is the only way out. Indeed, the very language with which this issue is debated clouds the issue. Terms such as "reconciliation with nature," or the "resurrection of nature," have mystical connotations which perhaps suggest that only a stubborn instrumentalism is a genuinely rational alternative.[68] Perhaps the issue is better discussed in terms of man's embeddedness in nature, his participation in nature. In any case, the anthropological examples discussed previously demonstrate that there exist combinations of attitudes toward nature that find no ready place in Habermas' system and yet that do not seem to represent a regression from the positive achievements of the Enlightenment (e.g., its "discovery" of individualism). In his eagerness to make sense of the dialectic of Enlightenment and to restore the cognitive dimension of critical theory, Habermas goes too far. He need not have rejected the blending of "basic attitudes" and the feedback of practice into attitudes, in order to rationally reinterpret the doctrine of mimesis in light of his communications philosophy. Indeed, perhaps the doctrine of mimesis need not be rationally reinterpreted at all. Perhaps the spontaneity and openness to experience—including humanity's experience of nature—that mimesis represents should be simply cherished, rather than utterly sublimated in human discourse. From this perspective, the problem is not so much that Habermas rejects McCarthy's "higher unity," but that his system lacks a certain imagination and openness to experience.

The utterly fascinating issue of what man's reconciliation with nature might look like cannot be further pursued here. It must suffice to state what it does not seem to require. This issue is best approached by again considering Habermas' evaluation of the *Dialectic of Enlightenment*. About this work Habermas states:

> This philosophy of history opens up a catastrophic view of a relation between spirit and nature that has been distorted beyond recognition. But we can speak of distortion only insofar as the original relation of spirit and nature is secretly conceived in such a way that the idea of truth is connected with that of a universal reconciliation—where reconciliation includes the interaction of human beings with nature, with animals, plants, and minerals.[69]

However, one is forced to ask why the goal of reconciliation with nature must be equated in the first place with a secret ambition for universal reconciliation with plants, rocks, and so forth? Apparently because Habermas continues to almost completely identify reconciliation with nature with the philosophy of history of Horkheimer and Adorno. "Psychic Thermidor and the Rebirth of Rebellious Subjectivity" (1980) is interesting in this regard. In this brief essay Habermas asks how Marcuse could retain his utopian optimism regarding the possibility of radical change in light of Marcuse's acceptance of the key thesis of the *Dialectic of Enlightenment*: that the progressive domination of nature requires that men and women progressively dominate their own natures. The answer, says Habermas (in what is really a quite standard reading of Marcuse's work) is Marcuse's postulation of a need for freedom, peace, and joy that is older than, and arises from below the levels of individuation and rationality.[70] Habermas argues that Adorno came to place his hope in aesthetics as a mode of thought that in its very speechlessness avoids the hegemony of conceptual thought.[71] It is such a strategy that Habermas characterizes as noncognitive. Marcuse, however, represents another alternative: the subject and object of reconciliation lie ready to be tapped in human nature itself, were the surplus repression of modern industrial society somehow overcome. Yet, even as Habermas is attracted by Marcuse's refusal to become trapped in the aporias of the dialectic of Enlightenment, Habermas in effect dismisses Marcuse's strategy with a single word. Marcuse, says Habermas, has a "chiliastic trust in a revitalizing dynamic of instincts."[72] Once more Habermas equates reconciliation with nature with eschatology.

Whether man's reconciliation with his own instinctual nature in the manner suggested by Marcuse is possible (or even desirable) is certainly open to question. Marcuse's interpretation is certainly incompatible with Freud's. In any case, this is a matter requiring a great deal of empirical study, in which anthropological research would certainly play a major role. However, the correctness of Marcuse's thesis is not at issue here.* The key point is that the notion of

* In this regard, it is one of the stronger arguments against Marcuse's thesis that he points to no other cultures that have taken even small steps toward his ideal of an eros-based society. That is, he points to no societies that come closer to his ideal than others (though Marcuse's brief reference to the Arapesh, in *Eros and Civilization*, should not be completely overlooked, pp. 197–98). Consequently, as Berndt and Reiche have argued, in "Die geschichtliche Dimension des Realitätsprinzips," one suspects that his ideal runs counter to some universal attributes of human nature (pp. 126–27). Marcuse might respond that humankind faces a world historically new situation, such that examples from other cultures would be irrelevant. Only modern industrial society is able to so reduce labor time that a society based upon play, and thus upon eros, becomes a theoretical possibility.

reconciliation with nature can be given cognitive content (unless cognitive content is somehow defined as excluding the instincts altogether) and that reconciliation need not be completely translated into a theory of communications in order to avoid the slippery slope that ends in universal reconciliation. Reconciliation with nature is given content and limited by a theory of human nature split off from its own true nature. Habermas, however, while recognizing that such a strategy differs from that of Horkheimer and Adorno, quickly goes on to equate Marcuse's program with theirs.* All three of these first-generation critical theorists are treated as eschatological utopians.

Why Habermas does not pursue this difference in strategy is unclear. Quite likely Habermas regards it as strictly a difference in degree when compared with his own fundamental reinterpretation of reconciliation in terms of a discourse ethic.[73] Habermas' motivation, though, is not the key point. The key point is that the concept of reconciliation need not be utterly sublimated in discourse (though this is not to say that mutual recognition in discourse is not a valid reinterpretation of an aspect of reconciliation) in order to avoid the aporias of the dialectic of Enlightenment. The anthropological examples discussed previously hint at other possibilities and combinations that might be realized.

However, once again the anthropological story is not so simple. As Stephen A. Marglin notes in a recent essay, it has become apparent to a number of anthropologists in the last several decades that many "primitive" societies in fact enjoy an enormous amount of leisure. Indeed, "one prominent anthropologist has labeled hunter-gatherers 'the original affluent society'" (Marglin, quoting from Marshall Stahlins' *Stone-Age Economics*, in *New York Review of Books*, 19 July 1984). If this is so, then it would not be inappropriate to look for aspects of Marcuse's ideal in such societies, and to regard its absence as evidence (though hardly definitive) against the possibility of his ideal.

*The discussion of the *Dialectic of Enlightenment* in the first chapter suggests that Habermas also pays insufficient attention to Horkheimer and Adorno's own understanding of reconciliation as reconciliation with the erotic and peaceful aspects of human nature that have come to be suppressed in the struggle for existence. That is, Horkheimer and Adorno are *also* concerned with humanity's reconciliation with its own nature, not merely with rocks, trees, and so forth. Is it possibly Habermas more than they who believes that reconciliation with nature cannot help but "secretly" imply universal reconciliation with all of nature?

Theoretical Pluralism and Dehumanization

Overcoming the theoretical impasse of the dialectic of Enlightenment is not the only reason that Habermas divides and categorizes knowledge as he does. Habermas also categorizes knowledge in order to prevent "dehumanization." Categorization, in this sense, is Habermas' attempt to deal with the same problem that the *Dialectic of Enlightenment* attempts to deal with via mimesis: the objectivation of mankind at the hands of instrumental reason. Habermas' approach is framed most dramatically in terms of the debate over the "displacement hypothesis." In fact, Habermas has recently framed the issue in these same terms in responding to Richard Rorty's *Philosophy and the Mirror of Nature*. Habermas states that Rorty seeks "to render the endangered humanity of the historical world compatible with the dehumanised reality of the objectivating sciences, which have been robbed of all human meaning."[74] However, Rorty's response to Habermas shows more clearly than ever before how many different issues were being mixed together in this debate over dehumanization. The entire issue cannot, however, be taken up here. To do so would require a thorough examination of the "linguistic turn" in analytic philosophy, as well as the theoretical pluralism of the new philosophy of science. I focus on only a facet of this issue, which directly pertains to Habermas' criticism of Rorty. Rorty sees himself as doing explicitly and precisely the opposite of what Habermas accuses him of doing: fostering the dehumanization of man. Rorty argues that he seeks to prevent all discourse from becoming normal discourse, in which the categories of knowledge are fixed and given. This, says Rorty, would indeed be dehumanization. "The resulting freezing-over of culture would be, in the eyes of edifying philosophers, the dehumanization of human beings. The edifying philosophers are thus agreeing with Lessing's choice of the infinite *striving for* truth over 'all of Truth.'"[75] I do not take up the issue of whether Rorty is correct or even whether his is the most fruitful way to formulate the issue. I rather address the question of why Habermas would see Rorty's project as doing precisely the opposite of what Rorty claims to be doing. It is this perspective that clarifies why there is a certain fixity in Habermas' categories.

Rorty argues against the notion, which he attributes to Habermas and Karl-Otto Apel, that we can avoid overconfident philosophical realism and scientific reductionism only by adopting a version of Kant's transcendental standpoint. What is needed is not Kant's epistemological distinction between the transcendental and the empirical standpoints, but rather his existentialist distinction between men and women as empirical selves and moral agents.[76] However, says Rorty, by tacitly identifying the moral agent with the constituting transcendental self, Kant left the road open to ever more complicated attempts to reduce freedom to nature, choice to knowledge.[77] This captures well, I think, the objections of McCarthy and especially Schmid against Habermas' theory of social evolution.[78] Rorty goes on to argue that he seeks to block "quasi-transcendental" strategies such as Habermas' by recasting ahistorical and permanent distinctions such as those between nature and spirit, objectivizing science and reflection, epistemology and hermeneutics, in terms of historical and temporary distinctions between the familiar and the unfamiliar, the normal and the abnormal. This way of treating these distinctions allows them to be seen not as dividing two areas of inquiry, but rather as the distinction between inquiry and something that is *not* inquiry, "but is rather the inchoate questioning out of which inquiries—new normal discourses—may (or may not) emerge."[79]

The only truth, says Rorty, to Habermas' claim that scientific inquiry is made possible, and limited, by the "inevitable subjective conditions" of inquiry, is that such inquiry is made possible by the adoption of practices of justification and that such practices have alternatives. However, says Rorty, these subjective conditions are in no sense usefully seen as inevitable ones discoverable by reflection upon the logic of inquiry. (It has been my argument that Habermas' more recent reconstructive science claims to grasp these "inevitable subjective conditions" almost as directly and immediately as reflection once did in his scheme.) Rorty adds that the inevitable subjective conditions are facts about what a given society or profession takes to be good grounds for assertions of a certain sort. Such disciplinary commitments are best studied, says Rorty, by the "usual empirical-cum-hermeneutic methods of 'cultural anthropology.'" From the point of view of the group in question, such subjective conditions are a combination of common sense and current theory and philosophy on the subject. From the point of view of the anthropologist, these conditions are empirical facts about the beliefs of a particular group. These are, says Rorty, incompatible points of view, in the sense that we cannot hold both viewpoints simultaneously. However, there is no reason to subsume them in a higher synthesis.[80]

In a paper presented at Bryn Mawr College, Habermas recently made a similar point. Indeed, it is important to grasp how far Habermas is willing to go to recognize the implications of relativism. In this paper Habermas suggests that he works within a tension or force field, as it were, between two conflicting viewpoints. On the one hand, he recognizes that any claim, when viewed from the vantage point of a third party, can be relativized, or historicized. On the other hand, for Habermas certain uses of language imply validity claims by their very nature (e.g., the "telos of truth" implicit in every speech act). Furthermore, this language use is dialogic: at least two persons are concerned with this validity claim, a concern that so involves them in it that they cannot at the same time treat it as relative. Only a madman can do both at once.[81] It is thus part of the human intellectual condition to make claims to truth, even as it is also a part of this condition to recognize that one can stand back and relativize any claim. Elsewhere Habermas puts it this way:

> In action oriented to reaching understanding, validity claims are "always already" implicitly raised. These universal claims . . . are set in the general structures of possible communication. In these validity claims communication theory can locate a gentle but obstinate, a never silent although seldom redeemed claim to reason, a claim that must be recognized de facto whenever and wherever there is to be consensual action.[82]

Habermas seems to reassert this "always already" claim in "A Reply," as Richard Bernstein points out in *Beyond Objectivism and Relativism*.[83]

Habermas' response here is not unlike his earlier response to Hesse, where he distinguishes between the *terminus a quo* and *terminus ad quem* of different types of knowledge in a way that seems to beg the issue of the claimed special status of reconstructions. That is, once relativism is admitted (even from a third party perspective), it is not clear that the assertion that communication oriented toward understanding always makes a "gentle but obstinate . . . claim to reason" possesses the universal force Habermas attributes to it. Under the assumption that third party relativism is a valid position,

Habermas' assertion would seem to apply only to those who accept (or whose culture tacitly accepts) the presumptions of Habermas' communicative ethic to begin with. To claim (in the absence of extensive empirical study) that all cultures necessarily make this tacit assumption would seem to involve Habermas (or his defender) in a return to some of Habermas' earliest transcendental claims about language. However, since Habermas' recent work has shown a desire to move away from such claims, his position is probably best seen as resting upon the empirical structure of communicative ethics as revealed by the theory of social evolution in the realm of language and morals.[84] In this light, and especially since the reconstructions of linguistic competence upon which this theory is based are, we have seen, no more than empirical hypotheses, Habermas' position would have to be seen as no more than a contingent empirical claim. Like the claimed special status of reconstructions (of which Habermas' "always already" claim is really a special instance), Habermas seems to want to treat as transcendental in its force what is really—according to his own response to his critics—merely an ordinary empirical hypothesis: that language is oriented toward rational agreement.

Perhaps Habermas would respond that he is not making a disguised transcendental claim, but merely a universal one. Such a response would be plausible, especially when one considers that if transcendental is understood as the study of the conditions of the very possibility of *x*, then there is no reason that these conditions cannot be empirically studied.[85] Such an approach—which substitutes empirical study for philosophical reflection—is sometimes known as transcendental realism. However, if this is to be Habermas' response, then once again the appropriate reply would seem to be that it is certainly the next step to actually go out and study these conditions empirically. As discussed previously, this has not been Habermas' approach, although it is indeed the case that he has drawn upon such studies, such as those of Piaget. The problem, as was also previously discussed, is that the studies Habermas draws upon are themselves hardly immune from the charge of cultural bias. Once again, Habermas seems to try to have it both ways. In this case, universal empirical claims are made in such a way that their contingency is downplayed by treating them at such a high level of abstraction that their empirical character is not immediately and readily recognizable.

All this is not to say that Rorty's own position is necessarily correct. The contradictions characteristic of a defense of the validity of relativism are well known and are perhaps not wholly avoided by Rorty. However, it is surprising how close Habermas comes to recognizing such a position. At the same time, it is apparent that Rorty's position is not adequately addressed by arguing that it risks dehumanizing humanity, and this is the primary issue at stake here. The fear of self-objectivation is, says Rorty, the fear that there will be objectively true or false answers to every question. Human worth would thus consist in knowing these truths, "and human virtue will be merely justified true belief." This is frightening, because it "cuts off the possibility of something new under the sun, of human life as poetic rather than merely contemplative." However, Rorty continues, humanity is not truly threatened by science or naturalistic philosophy. It is threatened by scarcity of food and by the secret police. "Given leisure and libraries, the conversation which Plato began will not end in self-objectivation—not because aspects of the world, or of human beings, escape being objects of scientific inquiry, but simply because free and leisured conversation generates abnormal discourse as the sparks fly upward."[86]

If the threat posed by Rorty's scheme is not really to endangered humanity, to what is it? It is that the project of grounding critical theory first upon reflection, then upon evolution in the realm of language and morals, will be shown to be jejune. It is that critical theory might be shown to have no special access to questions of emancipation and the good life. Habermas is still seeking to ground critical theory, all the while recognizing the limits of this project. However, recognizing these limits does not grant the right to subsequently ignore them. Nor does it seem quite fair to virtually equate the task of securing the autonomy of critical theory with that of preventing the dehumanization of humanity. Critical theory can continue to serve endangered humanity without making such a presumption.

The Neglected Virtues of Marcuse's Project

The criticisms against Habermas' scheme that I have reviewed in this chapter are powerful. Yet, it should not be overlooked that Habermas' work has proven enormously productive of fruitful debate and discussion. His responsiveness to criticism has contributed immeasurably to a dialogue in which the "sparks fly upward." Yet, if our standard of productive speculation is that it never "cuts off the possibility of something new under the sun," that it sees "human life as poetic rather than merely contemplative," then it is surprising that Marcuse's speculations have not received most positive attention of late, especially from those who are dissatisfied with the dualisms of Habermas' scheme.

Why has Marcuse's scheme not been more eagerly embraced? Why has even so-called sympathetic criticism, such as that collected in *Antworten auf Herbert Marcuse*, ostensibly a *Festschrift* on the occasion of Marcuse's seventieth birthday, been so unremittingly critical of Marcuse's program that Habermas was led, in his introduction, to call it an *Antifestschrift*? This question can be answered in several ways. Surely such analytic concepts as Marcuse's technological totality, which allows no opposition and thus leads the theorist into revolutionary pessimism or aesthetic retreat, are a large part of the story. That is, Marcuse does indeed run up against the barrier that Habermas tries to overcome with his dualistic concept of reason. For Marcuse, it seems as if liberation is an all or nothing affair. Even nature must cooperate. However, these theoretical objections are perhaps not the whole story. It is also the case that Marcuse strikes a deep chord. He speaks to a set of longings for peace, joy, satisfaction, and unity—with erotic self and world—which are challenging and threatening to rationalistic categories. Marcuse's is an underground political theory: a political theory that might be written by the erotic drives themselves, if they could communicate in sentences. The genre of work most similar to Marcuse's in many respects is that of the utopian authors: More, Bacon, St. Simon, Fourier.[87]

To suggest that Marcuse is the subject of excessively harsh criticism because he strikes a deep chord is, of course, too simple. One might say that there are unhealthy aspects to this chord. The peace Marcuse writes of is sometimes all too similar to the peace of the womb. Indeed, this is one reason why his new science is contradictory. To achieve Nirvana on earth requires that science and technology become enormously powerful and virtually autonomous instruments. However, there are other unhealthy aspects to Marcuse's vision of liberation. Marcuse argues that the dominant mythic cultures heroes are figures like Odysseus or Prometheus, clever tricksters who create culture at the price of perpetual pain. Against these heroes Marcuse sets the images of Orpheus and Narcissus.

> They have not become the culture-heroes of the Western world: theirs is the image of joy and fulfillment; the voice which does not command but sings; the gesture which offers and receives; the deed which is peace and ends the labor of conquest; the liberation from time which unites man with god, man with nature.[88]

Still, this is not the story of these antiheroes. Narcissus, it will be recalled, rejects the erotic charms of Echo (who could not initiate dialogue, but only reply) for the auto-eroticism of his own image. He finds his image so attractive that he pines away and dies while admiring it in the still water. Orpheus, Marcuse's other antihero, could charm wild beasts with his lyre. However, after striking a deal with Pluto to recover his wife Eurydice from Hades, Orpheus could not control his own desire and anxiety sufficiently to lead her back to this world. In violation of his bargain with Pluto, he seeks a reassuring glance of her, and she is snatched away from him forever. Thereafter Orpheus held himself apart from women, dwelling on his lost opportunity. Thracian maidens sought to captivate him, but he resisted their erotic charms, until one day they became so incensed that they drowned out the music of his lyre with their screams and tore him to pieces.[89] That Marcuse chooses Orpheus and Narcissus as his antiheroes, while virtually ignoring the fate of each (cf. *Eros*, p. 155), is revealing of the psychological dynamics of his vision of liberation. Is an erotic hero fixated on himself unto death really an image of fulfillment? Is an erotic hero who cannot control his own anxiety and longing sufficiently to reach safety, and in regret spends the rest of his life in mourning, rejecting eros utterly, an ideal? Surely the balance can be better struck than this.

If the unconscious chord that Marcuse strikes does not strike quite the right balance, it is nonetheless surprising that it has not found more resonance among theorists concerned with the fixed dualisms of Habermas' scheme. Whitebook poses modern biology as an example of a science not bound to the technical cognitive interest, and Stockman poses Rom Harré's scientific realism as an alternative.[90] Neither seems attracted to Marcuse's far more ambitious alternative, nor are McCarthy or Ottmann attracted.[91] However, if all our categories are tentative until further notice, then why not speculate about a science in which the aesthetic motive is blended with the theoretical? Habermas argues that "the loose ends of an objectivating natural science and of an interpretative approach to nature cannot be tied together."[92] However, our considerations suggest that the most that can be justifiably said is that it is not now apparent how these loose ends could be tied up. Concepts such as the "species interest" or the divergent evolutionary paths of science and language are potentially misleading, unless they are seen as pure description of the present intellectual division of labor. Given that this is the case, why should the philosophy of science—broadly interpreted to include the enterprises of Marcuse and Habermas—not concern itself with imagining how these loose ends could be spliced together? It would be silly, of course, to simply posit an aesthetic science and then go on to explore its implications. This is the stuff of utopian fiction. However, this is not Marcuse's project. *Eros and Civilization*, in conjunction with "On Hedonism," contains a quite elaborate explanation of why we lack an aesthetic science, the principles of human nature that would make one possible, and the social and economic changes that could possibly bring one about.

The question is not whether Marcuse or Habermas is correct. The answer to *this* question is surely that Habermas comes closer to describing the current place of science in society and philosophy, including its disconnection from aesthetics. The question is rather what is the purpose of grand philosophical speculation about the categories of human knowledge and action? There are, after all, alternatives to grand philosophical speculation. One alternative is to work out special problems within the normal discourse of a particular discipline, as the philosophy of science ordinarily does. Another alternative is to use the analytic tools of philosophy to illuminate pressing practical problems, such as why it is desirable to feed the hungry or curb the secret police. Still a third alternative is to test particular philosophical hypotheses against the evidence of this and other cultures. However, if one is determined to do grand speculation about science, then a certain theoretical playfulness, one that rejects old categories and creates striking new ones—an approach that sees the possibility of something new under the sun—seems appropriate. Grand speculation, whether it is Marcuse's or Habermas', is just that. That Habermas speculates about the givenness of the present order, whereas Marcuse speculates about the possibility of a future one, does not make Habermas' speculation less speculative. It just happens to better accord with the way things currently are.

The great virtue of the first generation of critical theorists is their refusal to equate the given, the currently existing, with what must be. I have called this refusal critical negativism. To be sure, this Hegelian inheritance has had its dark side. It reflects in part that "old spiritualist contempt for the finite and terrestrial world. . . . It is not a fight against particular socio-historical institutions (such as 'profit,' 'monopoly,' or even 'socialist bureaucracy'); it is a fight against objects and things."[93] Such an attitude is perhaps most dangerous when it leads to a contempt for reform and to partial solutions. Such contempt risks condemning actual men and women to this miserable reality until the millennium comes. Marcuse's project is not immune to criticism on these grounds. At the same time, much of Marcuse's project is pure speculation. So is Habermas'. Why should such speculation remain confined to categories drawn from the given? Or rather, why should such speculation remain confined to the *currently* given? Marcuse may or may not be correct that there exists an "originary synthesis" that ties all the loose ends, particularly those of subject and object, man and nature, together. This is, it will be recalled, his *Massstab*, or fundamental measure of the good life. However, it is certainly the case that groups of men and women have existed and currently exist who do not see nature under the horizon of the technical cognitive interest.[94] Furthermore, Habermas himself states that it was not until the science of Galileo that science was drawn under the technical cognitive interest.[95] Prior to this time, science and technology were loose ends to each other. Given these realities, is not speculation about new—or old—relationships between types of experience and knowledge a potentially fruitful undertaking?

Plato asks if his city of words "is any the worse drawn, then, because we can't show how it can be realized in fact?" (*Republic*, 472e). A traditional answer has been that it depends upon whether Plato's utopia embodies distinctions that are in accordance with nature. If, for example, there truly exists a natural division of labor that is somehow captured, even if in highly exaggerated form, by the myth of the metals, then Plato's can be a useful political challenge, goal, or standard. However, if his republic had depended upon the guardian class possessing the power of psychokinesis, for example, it could readily be dismissed as a piece of utopian fiction. Even though it might well express deep human needs and in this sense be in accord with nature, a republic based upon psychokinesis would not be in accord with nature as we know it. However, when what is natural or in accordance with nature is the question at stake to begin with, then our tolerance for speculation should be very high. Indeed, when such fundamental questions are in dispute, it is difficult to see how philosophy or social theory can justifiably reject outright a vision of an ideal world that does not compromise, that asks for the full amount of everything man has ever dreamed. The risk of holding to such a vision is despair or utopian putschism. The risk of ignoring it is not merely premature and unnecessary accommodation to the given. Ignoring it also risks driving the hopes and dreams that it represents underground or into the higher realm of idealist (or aesthetic) philosophy. This too is the dialectic of Enlightenment.

One might respond to this call for speculation about new categories of knowledge and experience that it has been tried and found wanting in Marcuse's case. Marcuse's project, like the critique of the dialectic of Enlightenment, has run into a theoretical dead end: it requires the resurrection of nature as its solution; i.e., it requires the philosophical equivalent of the eschaton.* This is a powerful argument and not to be dismissed. At the same time, I have suggested that Habermas' project of fixing and categorizing types of knowledge, holding them separate, may have reached an impasse. Originally intended to carve out a niche for critical theory and beleaguered humanity, safe from the imperialism of scientism, and later intended to justify the emancipatory commitment of critical theory, such a scheme has been shown by numerous sympathetic critics to require distinctions (e.g., between ordinary and reconstructive science) that cannot be upheld. Given that this seems to be the case, speculation based upon entirely new categories seems especially warranted. These new categories need not be Marcuse's. This is not my point. Rather, they should stem from free and playful thought *and should be recognized as such.* It is this recognition that keeps thought alive and the sparks flying upward.

The Legacy of the Enlightenment

The preceding considerations should not be read as suggesting that Marcuse's vision is so freely drawn as to be idiosyncratic or

*Whitebook makes the point that even to call Marcuse's program one of the "resurrection of nature" prejudices the discussion in advance. Though this term has become part of the jargon of the debate, and thus can hardly

merely part of an idealistic fight against objects and things. Indeed, in many respects Marcuse's scheme shares the understanding of science of the Enlightenment, in which this fight against things is material: the struggle against scarcity, toil, and sickness. (This is seen most clearly in Marcuse's 1933 article on labor.) It is Habermas who rejects the Enlightenment understanding of science more thoroughly than Marcuse. This is the case even though it is Habermas, not Marcuse, who adopts the Enlightenment ideal of "mature autonomy" [*Mündigkeit*] as his own. In "The Classical Doctrine of Politics," Habermas argues that the scientific approach to politics is characterized by its abstraction from every value but that of human survival. Every issue is quickly reduced to the lowest common denominator of whether it serves the survival of a particular group. The ancients focused on virtue, the moderns on survival. There is no common ground and no continuum between them.

> If the theoretically based point of departure of the Ancients was how human beings could comply practically with the natural order, then the practically assigned point of departure of the Moderns is how human beings could technically master the threatening evils of nature. To be sure, beyond the securing of mere survival, social philosophy is also concerned with the betterment, the easing, and the raising of life to a higher level. *Still, this is fundamentally different from the moral perfection of life*. The pragmatic forms of heightening the agreeableness and strength of life retain their reference to the positive, to the mere maintenance of life. *They still retain the comparative inflections of survival*, of surviving the elementary dangers to life: the physical threat posed by the enemy or by hunger.[96]

Habermas, it is apparent, sees a life of ease and happiness as the furthest implication of the scientific approach to politics. Yet Habermas does not consider a life of ease and happiness a very profound achievement, even should it be realized. "[Sir Thomas] More states this concept naively: 'What greater riches can there be than a life in

---

be avoided, it unfairly connotes that all attempts to go beyond the modern scientific approach to nature must be quasi-religious and mystical (Whitebook, "Nature and Habermas," p, 55, n. 38).

happiness and peace, with all cares removed, without being worried about one's own daily bread.'"[97]

What greater riches indeed, Marcuse might well reply, than a life in happiness and peace! It is Marcuse's appreciation of the relationship of science to this vision of the good life, an appreciation particularly characteristic of the utopian popularizers of science, Thomas More and Francis Bacon, which links Marcuse's understanding of science to that of the Enlightenment more closely than Habermas'.* For Habermas the higher "classical" values remain the preferred ones. Physical comfort and ease, even happiness, are valued by Habermas primarily as means. They contribute to the free time and roughly equal status among participants necessary for the unconstrained public discussion of value choices. The free discussion of alternative value choices, leading to agreement that may be characterized as substantively (communicatively) rational, is to be sure not an express goal value in Habermas' system. Yet, it becomes in some respects a goal by default. The substance of the good life remains a consideration beyond the boundaries of Habermas' system. His is ultimately a philosophical method by which we may attach the predicate "rational" or "irrational" to particular forms of social life. Which predicate is awarded depends almost entirely upon the attributes of the process by which social decisions are made—upon whether the decisions are made discursively, free of domination by social subsystems based upon money, power, and technical expertise.

For Habermas the purpose of social theory is to enunciate the conditions under which the good life may be decided upon rationally. This is how decisionism, and the hegemony of functionalist thought, is overcome: not by showing that a particular way of life is best, but by demonstrating what constitutes a genuinely and comprehensively rational decision-making process. Habermas avoids the intellectual rigidity, to say nothing of the incipient authoritarianism, of those who overcome decisionism only by positing the su-

* Bacon's account of the "Feast of the Family," in *New Atlantis*, seems to capture the profoundly physical, indeed erotic (The feast celebrates "any man that shall live to see thirty persons descended of his body alive together, and all above three years old.") character of the good life made possible by a society based upon organized scientific and technological research (pp. 59–63).

periority of a particular way of life.[98] The substance of the good life is beyond the pale of Habermas' social theory (except to the limited degree to which the formal outlines of the good life would mirror the conditions of discourse). The details of the good life become a task for art. In the last analysis, Habermas evaluates science almost exclusively in terms of the challenge it poses to communicative rationality. The idea that science could contribute to emancipation by illuminating the intellectual darkness is quite secondary.

For quite different reasons Marcuse rejects the idea that science may be emancipatory by lighting the intellectual darkness. Neither shares this aspect of the Enlightenment's understanding of science. Habermas focuses upon how language may light the darkness by speaking to the meaning of human existence. Marcuse believes that lighting the intellectual darkness is itself ultimately a detour from genuine emancipation. Yet, unlike Habermas, Marcuse draws science into his vision of emancipation. By embracing the contribution of science to "a life in happiness and peace," Marcuse partakes of the Enlightenment understanding of science. In fact, perhaps more thoroughly than any contemporary theorist ever has, Marcuse embraces precisely those aspects of modern thought that Habermas labels scientistic. Marcuse's goal is indeed the furthest extension of the search for survival: the classical utopian ideal of a life of happiness, peace, and contentment, with all cares removed, without having to worry about one's daily bread, that Habermas considers naive [*naiv*].[99] The difference between these authors is profound not merely because Marcuse values pleasure more highly than does Habermas. Rather, Marcuse values a "scientific" politics for precisely those reasons Habermas is averse to it. Scientific politics promises a life in which we no longer need be concerned with "higher" values. These values at best have been a detour from genuine gratification, at worst an ideological veil.

Science is valuable for Marcuse precisely because it does not partake of the richness of the classical concept of reason, including its appreciation of rhetoric. For Marcuse, scientific reason grasps things in their empirical immediacy. It ignores their history, their potential. Science is thus well suited to a world in which men and women would no longer debate the good life, but instead live a historically new form of it made possible by scientific and technological progress itself. Scientific reason, guided by Reason, one day renders Reason unnecessary. The one-dimensional character of science well suits it to the rationality principle that would prevail in an entirely different world—a world liberated from the need for rational discourse itself. Understood in this vein, Marcuse's vision of liberation does not require the "resurrection of nature." It requires the utter completion of the scientific project of the Enlightenment, the liberation of man from scarcity and toil. In such a world, the denial of self necessary to dominate nature would no longer be necessary. Such a denial is a characteristic of the incomplete scientific and technological transcendence of nature, not its fulfillment. In a world freed from scarcity and toil, a new science might indeed discover a new nature. However, Marcuse's vision of liberation (in this his dominant strategy; that his work is characterized by competing strategies was noted) does not seem to require it.*

Though Marcuse and Habermas share a common intellectual tradition, it is nonetheless difficult to imagine a vision of the good life more different from Habermas' than is Marcuse's. The irony of Marcuse's project is that it rests upon an industrial world view—that science and technology will make increasing amounts of ever-cheaper energy available, so that human labor can be progressively replaced by giant machines—that seems to have become obsolete just as Marcuse embraced it as the means of absolute freedom. However, this too is not necessarily given for all time. Surely, Habermas' skepticism regarding the contribution of science to human emancipation better fits the spirit of the postwar world. Yet, in Marcuse's audacious use of science as a vehicle of absolute freedom there is reflected an understanding of the origins of modern science in the Enlightenment that is absent in Habermas' work.

*The usual objection to saving Marcuse from himself, as it were, in this fashion is that in Marcuse's scheme surplus repression will not vanish until men and women have new needs. However, these new needs cannot emerge until surplus repression is banished (*Five Lectures*, p. 80). Such a vicious circle—new needs require new institutions, which can only be built on the basis of new needs—presumably led Marcuse into his so-called aesthetic retreat. Nevertheless, there is no reason to close the "technological totality" as tightly as Marcuse sometimes seems to. In fact, in *Counterrevolution and Revolt*, as well as "Protosocialism and Late Capitalism," to mention just two sources, Marcuse identifies a number of gaps within the system, within which new needs might develop and become a social force (*Counterrevolution*, pp. 129–34; "Protosocialism," pp. 41–46).

NOTES

Chapter 1: The Issues Involved

1. Habermas interprets his difference with Marcuse along precisely these lines. See "Technology and Science as 'Ideology,'" pp. 86–89. Thomas McCarthy, in *The Critical Theory of Jürgen Habermas*, accepts Habermas' formulation of the difference, pp. 20–22. McCarthy's book is hereafter cited as CTJH. Others, such as David Held, in *Introduction to Critical Theory: Horkheimer to Habermas*, appear to go along with this distinction, pp. 243–44; 304–7. Still others, however, seek to play down the radical character of Marcuse's new science. Morton Schoolman, in *The Imaginary Witness: The Critical Theory of Herbert Marcuse*, does so implicitly, pp. 282–86. William Leiss, in *The Domination of Nature*, argues explicitly that Marcuse's new science is not radical, pp. 199–212. See too Leiss' "Technological Rationality: Marcuse and his Critics." See also his "The Problem of Man and Nature in the Work of the Frankfurt School." Ben Agger contrasts the positions of Marcuse and Habermas on science. He generally agrees with the distinction in the text. He finds Marcuse's position preferable because it is more radical. See "Marcuse and Habermas on New Science." Douglas Kellner, in *Marcuse and the Crisis of Marxism*, agrees with Agger (p. 332), but then seems to argue that what he calls Marcuse's "new technology" is radical only insofar as the uses to which it is put are concerned (pp. 330–38). Kellner's recent book is most comprehensive; nothing else comes close in this regard. Norman Stockman, in "Habermas, Marcuse, and the *Aufhebung* of Science and Technology," seems to generally agree with the distinction drawn in the text.

2. 2 vols. (Frankfurt a.M.: Suhrkamp, 1981); hereafter cited as *Theorie*.

3. Joel Whitebook, "The Problem of Nature in Habermas." Thomas McCarthy, "Rationality and Relativism: Habermas' 'Overcoming' of Herme-

neutics," pp. 75–78. Henning Ottmann, "Cognitive Interests and Self-Reflection," pp. 87–92. "Will-to-power" is Ottmann's phrase.

4. McCarthy, "Rationality and Relativism," p. 77. CTJH, pp. 110–36.

5. Karl-Otto Apel, *Towards a Transformation of Philosophy*. Albrecht Wellmer, *Critical Theory of Society*.

6. Herbert Marcuse, "Industrialization and Capitalism in the Work of Max Weber." Jürgen Habermas, "Technology and Science as 'Ideology.'"

7. Marcuse, "Industrialization," pp. 225–26. Habermas, "Technology and Science," pp. 86–87.

8. Habermas, "Technology and Science," pp. 86–87.

9. Ibid., p. 86.

10. Ibid., pp. 87–88.

11. These refinements are begun in Habermas' "A Postscript to *Knowledge and Human Interests*."

12. Jürgen Habermas, "A Reply to my Critics," pp. 243–45.

13. Habermas, "Some Difficulties in the Attempt to Link Theory and Practice," p. 14.

14. Ottmann, "Cognitive Interests," p. 89.

15. Habermas, "A Reply," p. 274.

16. *Conjectures and Refutations*, p. 111.

17. Ibid., p. 110.

18. Ibid.

19. Ibid., pp. 111–13.

20. Habermas, "A Positivistically Bisected Rationalism," p. 208.

21. Francis Bacon, *The Great Instauration* and *New Atlantis*. The preface to *The Great Instauration*, which was to be a program by which human knowledge would be made relevant to the betterment of the human condition, shows this clearly, pp. 7–17.

22. Max Horkheimer and Theodor Adorno, *Dialectic of Enlightenment*, p. 57.

23. Whitebook, "Nature in Habermas," p. 42.

24. Ibid., p. 44. Habermas agrees with Whitebook's interpretation in "A Reply," p. 240.

25. Conant, *Science and Common Sense*, quoted in *Human Understanding*, vol. 1, by Stephen Toulmin, p. 372; hereafter cited as *Understanding*.

26. Boris M. Hessen, "The Social and Economic Roots of Newton's *Principia*," quoted in Toulmin, *Understanding*, p. 303.

27. Ashton, *The Industrial Revolution, 1760–1830*, quoted in Thomas Kuhn, *The Essential Tension*, p. 141.

28. Mousnier, *Progrès scientifique et technique au XVIIIe siècle,* cited in Kuhn, *The Essential Tension*, p. 141.

29. Kuhn, *The Essential Tension*, p. 142.

30. Habermas, *Toward a Rational Society*, pp. 53–54.

31. Toulmin, *Understanding*, p. 370.

32. *Theorie* 1: 533. Translation from a review of *Theorie* by Johannes

Berger. The first volume of *Theorie* has recently been translated as *The Theory of Communicative Action: Reason and the Rationalization of Society*, trans. Thomas McCarthy; hereafter cited as *Theory*. Page references will also be given to the translation, when appropriate. In this case the reference to the translation is pp. 398–99.

33. The term dialectic of Enlightenment refers to Horkheimer and Adorno's thesis; *Dialectic of Enlightenment* is the title of their book.

34. Horkheimer's *Eclipse of Reason* contains by far the clearest exposition of the Frankfurt School's critique of reason.

35. Esp. pp. 223–33.

36. "Traditional and Critical Theory," pp. 239–43. Barry Kātz, in *Herbert Marcuse and the Art of Liberation*, p. 127, n. 30, captures critical theory's understanding of positivism well.

37. *Dialectic of Enlightenment*, pp. 20–42.

38. Ibid., pp. 43–80.

39. Held, *Introduction to Critical Theory*, p. 404.

40. Erich Fromm's *Escape From Freedom* is an excellent account of this process. Fromm was a member of the Frankfurt School during its early years.

41. Adorno, "Subject and Object," p. 499. Quoted by Martin Jay, in *Adorno*, pp. 63–64.

42. Jay, *Adorno*, p. 64.

43. *Dialectic of Enlightenment*, p. 40. Jay, *Dialectical Imagination*, p. 267. Habermas' sophisticated discussion and criticism of what reconciliation means to Adorno, especially, is addressed in chapter 9.

44. Horkheimer, *Eclipse of Reason*, p. 127.

45. Ibid., pp. 77–78.

46. *Theorie*, esp. vol. 1, sec. 4.

Chapter 2: Freedom and Labor in Marcuse's Early Works

1. See, for example, Robert D'Amico's review of William Leiss' *The Domination of Nature*, and Leiss' response. Leiss argues that the attribution of new science tendencies to Marcuse is the result of taking a few quotes out of context. D'Amico responds with more quotes.

2. Kātz, *Herbert Marcuse and the Art of Liberation*, pp. 35–108.

3. Marcuse, *Der deutsche Künstlerroman*, in *Herbert Marcuse: Schriften*, 1: 10–11. (There are now several volumes in print of what apparently will not be a complete collection of Marcuse's works). Kātz, *The Art of Liberation*, p. 47.

4. *Künstlerroman*, p. 10.

5. Ibid., pp. 332–33; my translation. I translate *Lebensunfähigkeit* as "unworldliness," because this seems to best capture Marcuse's sense, though it is hardly a literal translation.

6. Ibid., p. 16; quoted and translated by Kātz.
7. Kātz, *The Art of Liberation*, p. 50. *Künstlerroman*, p. 307.
8. *Künstlerroman*, p. 326.
9. Ibid.; quoted and translated by Kātz, p. 51.
10. Ibid., p. 330. I have found Kellner's discussion of this issue (pp. 28–30) most helpful, but I have altered his translation.
11. Marcuse, *The Aesthetic Dimension*, p. 69.
12. *Künstlerroman*, p. 193; quoted and translated by Kellner, p. 26.
13. Kātz, *The Art of Liberation*, p. 82.
14. Marcuse, *Hegels Ontologie und die Theorie der Geschichtlichkeit*, p. 303 (the title of the 2d and 3d editions is slightly different; cf. p. 22, this text); Kātz, *The Art of Liberation*, p. 82.
15. See, for example, Herbert Marcuse, Jürgen Habermas, Heinz Lubasz, Tilman Spengler, "Theory and Politics: A Discussion," in *Telos*, no. 38 (1978–1979). Habermas' "Psychic Thermidor and the Rebirth of Rebellious Subjectivity," written shortly after Marcuse's death, is perhaps their most fruitful "exchange." It is considered later.
16. Kātz, *The Art of Liberation*, p. 83. *Hegels Ontologie*, pp. 84–85.
17. Kātz, *The Art of Liberation*, p. 83.
18. Jean-Michel Palmier, in *Herbert Marcuse et la nouvelle gauche* (Paris, 1973), maintains that a "decisive rupture" separates *Hegels Ontologie* from the rest of Marcuse's work (Palmier, p. 91; Kātz, p. 81, n. 67).
19. Wisengrund-Adorno, "Besprechung von Herbert Marcuses *Hegels Ontologie*," in *Zeitschrift für Sozialforschung* 1 (1932): 409; quoted by Kātz, *The Art of Liberation*, p. 84. Kellner, *Marcuse and Marxism*, p. 76.
20. *Hegels Ontologie*, p. 236; quoted and translated by Kātz, p. 83.
21. *Telos*, no translator given. The German original is included in vol. 1 of *Marcuse: Schriften*.
22. Ibid., p. 26.
23. Alfred Schmidt, "Existential-Ontologie und historischer Materialismus bei Herbert Marcuse," p. 44.
24. "Contributions," p. 34. I have employed my translation from the original "Beiträge zu einer Phänomenologie des Historischen Materialismus," p. 68. The unattributed translation in *Telos* renders "die *Existenz* geschichtlichen Daseins" simply as "history"!
25. Schmidt, "Existential-Ontologie," p. 27.
26. Schoolman, *Imaginary Witness*, pp. 10–12.
27. Ibid., p. 12.
28. Morton Schoolman, in "Introduction to Marcuse's 'On the Problem of the Dialectic,'" in *Telos*, no. 27 (Spring 1976), cites seven works that address Heidegger's influence on Marcuse, p. 3, n. 2.
29. "Contributions," pp. 21–23.
30. Ibid., p. 23.
31. Schmidt, "Existential-Ontologie," p. 47.

32. *Telos,* p. 22. The German original is included in *Marcuse: Schriften,* vol. 1.
33. This essay is included in *Marcuse: Schriften,* vol. 1.
34. Ibid., p. 359. See too *Hegels Ontologie,* chap. 26.
35. Ibid., pp. 357–60.
36. Ibid., p. 361; my translation.
37. Ibid., pp. 363–65.
38. Heidegger, *Platons Lehre von der Wahrheit;* quoted in Alfred Schmidt, *The Concept of Nature in Marx,* p. 234.
39. Schmidt, "Existential-Ontologie," pp. 22–24.
40. Marcuse, *One-Dimensional Man,* pp. 236–37.
41. *Telos,* pp. 14–17.
42. Ibid., pp. 14–15. I have altered the translation slightly. Cf. "Über die philosophischen Grundlagen des wirtschaftswissenschaftlichen Arbeitsbegriffs," p. 16. This essay is also included in *Marcuse: Schriften,* vol. 1.
43. "The Concept of Labor," p. 17.
44. Schmidt, *The Concept of Nature in Marx,* pp. 76–93.
45. "The Concept of Labor," p. 25.
46. Ibid., pp. 29–32.
47. Ibid., p. 15. I have altered the translation slightly.
48. "The End of Utopia," p. 63.
49. Kellner, *Marcuse and the Crisis of Marxism,* p. 89.
50. Ibid.
51. Lucio Colletti, *From Rousseau to Lenin,* p. 130. Colletti is referring to Marcuse.
52. *Hegels Ontologie,* chaps. 13–14, esp. pp. 162–66. *Reason and Revolution,* p. 163.
53. *One-Dimensional Man,* pp. 232–34.
54. Schmidt, *The Concept of Nature in Marx,* p. 163. The internal quote is from Walter Benjamin.
55. *An Essay on Liberation,* p. 22. See too *Counterrevolution and Revolt,* p. 65.

Chapter 3: The Ground of Absolute Freedom in Eros

1. Malinovich cites Harold Laski regarding the character of freedom in the eighteenth and nineteenth centuries.
2. Malinovich, pp. 158–59.
3. Ibid.
4. Heide Berndt and Reimut Reiche, "Die geschichtliche Dimension des Realitätsprinzips," pp. 107–10. Schoolman, in *Imaginary Witness,* focuses upon the biological dimension of sublimation according to Freud, pp. 92–

108. This discussion subsumes his several articles in *Telos* on Marcuse's psychological theory.

5. *Eros and Civilization*, p. 76.

6. Sigmund Freud, *Civilization and its Discontents*, pp. 36–37.

7. Ibid., esp. pp. 91–104.

8. This is recognized by Schoolman, esp. pp. 102–17, and Berndt and Reichc, pp. 105–10.

9. *Eros*, pp. 27, 98–99.

10. Ibid., p. 27.

11. Ibid., pp. 78–79.

12. Ibid., p. 79.

13. "The Obsolescence of the Freudian Concept of Man," pp. 56–58.

14. *Eros*, p. 142.

15. Ibid., p. 178.

16. Ibid., pp. 193–95. See also *An Essay on Liberation*, pp. 88, 91.

17. *Eros*, p. 195.

18. Ibid., p. 196.

19. Ibid., p. 179.

20. Anthony Giddens, "The Improbable Guru: Re-Reading Marcuse," p. 161.

21. *An Essay on Liberation*, p. 31.

22. *Eros*, p. 113.

23. Ibid.

24. Ibid., p. 173. See also *Counterrevolution*, p. 66.

25. *One-Dimensional Man*, p. 236.

26. "Philosophy and Critical Theory," in *Negations*, p. 146. I have altered the translation in order to render *Bestehenden* as "existent," not "the powers that be." Cf. the original, "Philosophie und kritische Theorie," p. 114.

27. Frank Manuel and Fritzie Manuel, *Utopian Thought in the Western World*, pp. 122–25; 253–55.

28. Lasch, *The Minimal Self*, p. 233. Lasch relies heavily on Melanie Klein, especially the essays collected in her *Envy and Gratitude and Other Works, 1946–1963* (New York: Delacorte Press, 1975), in support of this claim.

29. Lasch, *Minimal Self*, p. 235. Marcuse's criticism of Brown, in "Love Mystified: A Critique of Norman O. Brown," addresses only his later and more mystical *Love's Body* (New York: Vintage Books, 1966), not his earlier *Life Against Death* (Middletown, Connecticut: Wesleyan University Press, 1959).

30. Lasch, *Minimal Self*, p. 177.

31. Ibid., pp. 226–27; 253–259.

32. *Eros*, pp. 146–49.

33. Lasch, *Minimal Self*, p. 184.

34. *Eros*, pp. 215–16.

35. Lasch, *Minimal Self*, pp. 177, 247.

Chapter 4: Marcuse's New Science and its Dissolution in Freedom

1. Kātz, *The Art of Liberation*, p. 162.
2. "On Science and Phenomenology," pp. 286–87; my emphasis. Marcuse's "Some Social Implications of Modern Technology," in *Studies in Philosophy and Social Science* (1941), touches on a number of issues discussed in this chapter. However, it is a synthesis of sorts. Each point finds its origins, as well as its complete development, in other works. I focus on these other works.
3. *One-Dimensional Man*, pp. 157–58.
4. Marcuse, "Struggle Against Liberalism in the Totalitarian View of the State," pp. 5–6.
5. In *Five Lectures: Psychoanalysis, Politics, and Utopia*, pp. 44–61.
6. Ibid., p. 66.
7. *One-Dimensional Man*, pp. 157–58.
8. Ibid., pp. 133, 137, 140.
9. MacIntyre, *Marcuse*, p. 95.
10. Aron Gurwitsch, "Comment on the Paper by H. Marcuse," p. 298.
11. Ibid., p. 301; 303–4.
12. *One-Dimensional Man*, pp. 146–52.
13. "On Science and Phenomenology," pp. 284–86.
14. *One-Dimensional Man*, pp. 156–59.
15. Kātz, p. 39.
16. Edmund Husserl, *The Crisis of European Sciences and Transcendental Phenomenology*, pp. 3–18.
17. Ibid., esp. pp. 178–90.
18. Aron Gurwitsch, "Comment on the Paper by H. Marcuse," pp. 303–4.
19. "On Science and Phenomenology," p. 287.
20. *One-Dimensional Man*, p. 140.
21. *Negations*, pp. 222–24.
22. *One-Dimensional Man*, pp. 160–62. It is this particular interpretation of Husserl that Gurwitsch criticizes, pp. 303–4.
23. *The Nicomachean Ethics*, 1097a15–1097b22; 1177b33–1179b7, and throughout books 2 and 10 generally.
24. "On Hedonism," p. 161.
25. Ibid., p. 171.
26. Ibid.
27. Schoolman, *Imaginary Witness*, p. 123.
28. *An Essay on Liberation*, p. 32.
29. Ibid.
30. "On Hedonism," p. 177.
31. Ibid.

32. See Schmidt, *The Concept of Nature in Marx*, pp. 51–61, and the appendix.
33. "On Hedonism," p. 198.
34. *One-Dimensional Man*, pp. 166–67.
35. Ibid.
36. *An Essay on Liberation*, pp. 29–31.
37. Kuhn, *Essential Tension*, pp. 115–18.
38. Several contemporary philosophers of science, such as Paul Feyerabend, are certainly more radical than Marcuse on this point. Cf. pp. 83–86, this text.
39. Marcuse's "The Responsibility of Science" is an excellent brief study of the way in which social priorities (e.g., the "warfare state") affect the direction of scientific progress. Criticism of science, as this essay demonstrates, can be radical indeed without invoking a radical epistemology, such as the "new sensibility."
40. *One-Dimensional Man*, p. 166. Kellner, in *Marcuse and Marxism*, shows how central this concept is to Marcuse's project (pp. 339–53).
41. *Hegels Ontologie*, p. 236; Kātz's translation.
42. *An Essay on Liberation*, p. 31.
43. *Counterrevolution and Revolt*, pp. 68–69.
44. "Sartre's Existentialism," p. 175.
45. *The Aesthetic Dimension*, esp. pp. 1–21.
46. Freedom and labor are separated utterly in *Eros and Civilization*, 1954. The abolition of labor is the centerpiece of man's triumph over nature in *One-Dimensional Man*, 1964, p. 179.
47. Schoolman, *Imaginary Witness*, pp. 324–49.
48. *Counterrevolution*, p. 99.
49. Ibid., p. 111. Habermas disagrees with this "in the last analysis" interpretation of Horkheimer and Adorno's position, as will be shown.
50. *Aesthetic Dimension*, pp. 58–59; my emphasis.
51. Ibid., p. 69.
52. Ibid., p. 59; "Nevertheless, it was so beautiful!"
53. The essay is subtitled "Toward a Theoretical Synthesis Based on Bahro's Analysis."
54. Bahro's *Die Alternative* originally appeared in 1977.
55. "Protosocialism," p. 30. Marcuse does not elaborate. Freedom in necessity, it appears, remains but a beautiful promise.
56. Ibid.
57. Ibid., pp. 43, 46.
58. Ibid., pp. 46–47. Marcuse tacitly but clearly rejects the centralist tendencies in Bahro's analysis, such as Bahro's overestimate of the emancipatory function of the "collective intellectual."
59. "Lord Bacon," *Critical and Historical Essays*, p. 212.

Chapter 5: Habermas: Science and Survival

1. "Wozu noch Philosophie?", p. 34; my translation. This collection of essays has recently been translated as *Philosophical-Political Profiles*. A translation of "Wozu noch Philosophie?", rendered as "Why More Philosophy?", can also be found in *Social Research*.
2. *Theory and Practice*, pp. 44–81.
3. Ibid., p. 42.
4. Leo Strauss, "What is Political Philosophy?", pp. 39–57.
5. "Classical Doctrine of Politics," p. 42.
6. Strauss, "Plato," pp. 236–37. Karl Popper, *The Open Society and its Enemies*, esp. chaps. 3, 9.
7. "Classical Doctrine," pp. 57–58.
8. Ibid., pp. 54–55.
9. Ibid., my emphasis. The translation is faulty, for it translates a phrase used once in the original, "*rucken . . . auf eine Ebene*," in two different ways consecutively. Cf. "Die klassische Lehre von der Politik in ihrem Verhältnis zur Sozialphilosophie," p. 60.
10. "The Classical Doctrine" was delivered as an *Antrittsvorlesung* in 1961, and was first published in 1963.
11. "Classical Doctrine," pp. 46, 74.
12. Ibid., p. 44. The reference to Popper's *Logik der Forschung* is readily missed in the translation, which renders *Forschung* as *investigation*, a literal but misleading translation. Cf. "Die klassische Lehre," p. 50.
13. "Classical Doctrine," p. 74. Habermas' phrase is "*einsam und stumm*." Cf. "Die klassische Lehre," p. 79.
14. "Classical Doctrine," p. 60.
15. Hannah Arendt, *The Human Condition*, p. 301. Habermas' only full-length study of Arendt does not treat her view of science and technology. See "Hannah Arendts Begriff der Macht." Translated as "Hannah Arendt's Communications Concept of Power."
16. *Human Condition*, pp. 257–68; 289–95.
17. Ibid., pp. 301–2.
18. See John Viertel's "Translator's Note on German Terms," in *Theory and Practice*, pp. vii–viii.
19. Habermas, *Zur Logik der Sozialwissenschaften*, pp. 327–28.
20. Habermas, "A Positivistically Bisected Rationalism," p. 222.
21. "Classical Doctrine," p. 75. The internal quote is from Vico.
22. See Habermas' essays on Adorno, "Theodor W. Adorno: Ein philosophierender Intellektueller" and "Theodor W. Adorno: Urgeschichte der Subjektivität und verwilderte Selbstbehauptung." See esp. pp. 181–84. Also in the collection *Philosophical-Political Profiles*. See too "Dogmatism, Reason and Decision: On Theory and Praxis in Our Scientific Civilization," p. 267.

23. Thomas Hobbes, *Leviathan*, ed. C. B. Macpherson (Harmondsworth, England: Penguin Books, 1968), esp. chaps. 13–14.
24. Whitebook, "Nature in Habermas," p. 44.
25. Anthony Giddens, in "Labor and Interaction," cites an interview with Habermas by Detlef Korster and Willem van Reijen as his source for this claim, p. 301, n. 9.
26. *Knowledge and Human Interests*, chaps. 4–6.
27. Mary Hesse, "Science and Objectivity," p. 98.
28. Habermas, "A Reply," p. 274.
29. See Habermas' *Theorie der Gesellschaft oder Sozialtechnologie—was leistet die Systemforschung*?, written as a debate with Niklas Luhmann.
30. In *Theory and Practice*, pp. 142–69. I have found Giddens' brief discussion of this essay in "Labor and Interaction" very helpful, and I frequently refer to it.
31. Habermas, "Labor and Interaction," p. 159.
32. Ibid., p. 163.
33. Ibid., p. 168.
34. In chap. 11 of *Knowledge*, Habermas argues in a similar fashion that Freud "scientistically" misunderstood his own project: he thought it was a science, whereas it was actually an interpretative (hermeneutic) discipline.
35. *Knowledge*, p. 45; quoted by Giddens, p. 151.
36. *Knowledge*, pp. 61–63.
37. Albrecht Wellmer, in *Critical Theory of Society*, develops the criticism of the latent positivism in Marx even more aggressively than Habermas does. See esp. chap. 2.
38. "Technology and Science," p. 87.
39. Ibid., p. 99. The German term is *technisch verwertbares Wissen*. See "Technik und Wissenschaft als 'Ideologie,'" p. 73.
40. *Knowledge*, pp. 308–9.
41. See "Technology and Science," p. 87.
42. McCarthy, "Rationality and Relativism," p. 62.
43. *Knowledge*, p. 196.
44. "Some Difficulties in the Attempt to Link Theory and Praxis," p. 8.
45. McCarthy, "Rationality and Relativism," p. 64.
46. "A Postscript," p. 174.
47. *Knowledge*, pp. 127–29.
48. "A Reply," pp. 244–45.
49. Ibid.
50. Hesse, "Science and Objectivity," p. 102.
51. "A Reply," p. 277.
52. Marx Wartofsky, *Conceptual Foundations of Scientific Thought*, p. 253.
53. Ibid., pp. 286–87.
54. Ibid., p. 253.
55. "A Reply," p. 278.
56. Wartofsky, p. 252.

57. Ernest Nagel, *The Structure of Science*, p. 338.
58. Ibid., pp. 338–39.
59. Ibid., p. 340.
60. Keat, pp. 118–28.
61. Feyerabend, *Philosophical Papers*, 1: 91–94.
62. Feyerabend, "Against Method," pp. 81–91. *Philosophical Papers*, 1: x–xi.
63. Harold Brown, "Incommensurability."
64. Brown, p. 3. Kuhn, "Theory Change as Structure Change," pp. 300–301.
65. Kuhn, "Theory Change as Structure Change," pp. 300–301.
66. Rorty, "In Defense of Eliminative Materialism." Bernstein, *Praxis and Action*, p. 284.
67. Feyerabend, *Against Method*, chap. 2.
68. Ibid., p. 30.
69. Imre Lakatos, "Falsification and the Methodology of Scientific Research Programmes," p. 184.
70. Hesse, "Science and Objectivity," p. 98.
71. For some thoughts on this issue see Feyerabend's *Science in a Free Society*, pp. 79–86, and Richard Rorty's *Philosophy and the Mirror of Nature*, pp. 306–11; 373–79.
72. Nagel, *Structure of Science*, pp. 129–30.
73. Hesse, "Science and Objectivity," pp. 112–15.

Chapter 6: Habermas' Early Studies of Science and the Emergence of Language

1. "Peirce's Logic of Inquiry," pp. 111–12.
2. Ibid., p. 112.
3. "The Self-Reflection of the Natural Sciences," pp. 124–25.
4. Ibid., p. 125.
5. Ibid., pp. 124–25.
6. "Theodor Adorno," pp. 181–86.
7. "The Analytical Theory of Science and Dialectics," p. 154.
8. *Zur Logik der Sozialwissenschaften*, p. 32.
9. "A Positivistically Bisected Rationalism," p. 208.
10. See chap. 1, pp. 10–11, this text.
11. "The Self-Reflection of the Natural Sciences," pp. 134–35. See too "Reason and Interest," p. 194; both in *Knowledge*.
12. Krüger, "Überlegungen zum Verhältnis wissenschaftlicher Erkenntnis und gesellschaftlicher Interessen," p. 212.
13. "A Positivistically Bisected Rationalism," p. 201.
14. Karl-Otto Apel, "Wissenschaft als Emanzipation?", p. 334.

15. The argument does not appear in Habermas' work in quite this form.
16. "The Self-Reflection," p. 135.
17. "Zu Nietzsches Erkenntnistheorie (ein Nachwort)," p. 262.
18. "Knowledge and Interests," in *Knowledge*, p. 314.
19. Albert, "Hermeneutik und Realwissenschaft," pp. 130–32.
20. "Comte and Mach," in *Knowledge*, p. 89.
21. Ibid.
22. Krüger points this out, pp. 208–9.
23. Maurer, "Jürgen Habermas' Aufhebung der Philosophie," p. 46.
24. Ibid.; my translation.
25. "The Self-Reflection," in *Knowledge*, p. 130.
26. Nikolaus Lobkowicz, in "Interest and Objectivity," points out the ambiguity of the diamond example, pp. 202–5. I follow some of his argument.
27. "Some Difficulties in the Attempt to Link Theory and Praxis," p. 14.
28. "A Postscript," p. 175. However, it should be noted that in *Theorie* Habermas once again seems to assert the close relationship between labor, experiment, and science. See *Theorie* 1: 297–98. *Theory*, p. 215.
29. Ibid., p. 180.
30. I alter Hesse's language a great deal in order to better fit her analysis to the present argument.
31. Hesse, pp. 103–4.
32. Habermas, "A Reply," p. 275.
33. Habermas, "A Postscript," pp. 180–82.
34. Popper, "Three Views Concerning Human Knowledge," pp. 107–14. Russell Keat, in *The Politics of Social Theory*, agrees that Habermas' interpretation of science is not instrumental. Keat, however, believes that Habermas should adopt an instrumental view of science (pp. 70–78).
35. "Theorie der Gesellschaft," p. 237; my translation.
36. "A Postscript," p. 161.
37. "A Postscript," p. 162.
38. P. 284.
39. Feyerabend, "Against Method."
40. *Against Method*, p. 30.
41. "A Postscript," p. 162. Habermas does not abandon this view. In "A Reply" (1982), Habermas states that Richard Rorty, in *Philosophy and the Mirror of Nature*, seeks "to render the endangered humanity of the historical world compatible with the dehumanised reality of the objectivating sciences, which have been robbed of all human meaning" (p. 277). This charge is discussed in detail in chapter 9.
42. Pp. 238–50 in "A Reply" show how much of this original view Habermas maintains.
43. *Legitimation Crisis*, pp. 112–16.
44. Keat, p. 102.

45. "Vorbereitende Bemerkungen zu einer Theorie der kommunikativen Kompetenz," pp. 107–8.
46. "Der Universalitätsanspruch der Hermeneutik," pp. 270–76.
47. Ibid., p. 274 ("sits on top of"), and p. 290 ("a genuine integration").
48. Ibid., p. 274; my translation.
49. "Zu Nietzsches Erkenntnistheorie," pp. 251–53.
50. Ibid., p. 257.
51. Ibid.: "As if truth were proven by the continued existence of man!"
52. Ibid., p. 260. Habermas reasserts this claim in "A Reply," p. 276.
53. Ibid., p. 259.
54. "Wahrheitstheorien."
55. In *Proceedings of the 1978 Biennial Meeting of the Philosophy of Science Association*, vol. 2.
56. McCarthy, "A Theory of Communicative Competence," pp. 148–54. See too McCarthy's CTJH, pp. 299–310, esp. p. 303.
57. Alan R. White, "Coherence Theory of Truth."
58. Maurer, "Habermas' Aufhebung der Philosophie," pp. 43–49.
59. Jean Piaget's *Genetic Epistemology* juxtaposes the way children learn to think with the most advanced intellectual constructs in order to make this point.

Chapter 7: Epistemology or Politics?

1. "The Self-Reflection," p. 137.
2. Lobkowicz, "Interest and Objectivity," p. 204.
3. "The Self-Reflection," p. 336, n. 30.
4. "Some Difficulties in the Attempt to Link Theory and Praxis," p. 3.
5. "Technical Progress and the Social Life-World," p. 52.
6. "The Scientization of Politics and Public Opinion," pp. 70–74. Shapiro translates *Können* as "potential"; I translate it as "ability." The German phrase is "die Dialektik von aufgeklärtem Wollen und selbstbewusstem Können." Cf. "Verwissenschaftlichte Politik und öffentliche Meinung," p. 135.
7. Dieter Senghaas points this out in "Sozialkybernetik und Herrschaft," p. 198.
8. "The Scientization of Politics," pp. 72–73.
9. "Technical Progress," p. 52.
10. Ibid., p. 57.
11. Ibid.
12. Ibid., p. 60. I have altered the translation slightly. See also "The Scientization of Politics," pp. 72–73.
13. "Technology and Science as 'Ideology,'" p. 113.
14. "Technical Progress," p. 59.
15. McCarthy, CTJH, pp. 377–80.

16. "The Scientization of Politics," pp. 70–72.

17. Ibid., pp. 67–68.

18. "Öffentlichkeit," p. 68. Translated as "The Public Sphere: An Encyclopaedia Article." This is a brief statement of the thesis of Habermas' *Habilitationsschrift, Strukturwandel der Öffentlichkeit.*

19. *Science, Technology, and Human Values*, 5 (Winter 1980): 43–44.

20. Joel Primack and Frank von Hippel, *Advice and Dissent: Scientists in the Political Arena*, pp. 257–58.

21. "Theorie der Gesellschaft oder Sozialtechnologie?", pp. 162–63.

22. *Advice and Dissent*, pp. 14–16.

23. *Telos*, p. 101. This essay is an abridged English version of "Können komplexe Gesellschaften eine vernünftige Identität ausbilden?"

24. For instance: Grant McConnell, *Private Power and American Democracy*; Theodor Lowi *The End of Liberalism*; Louis Hartz, *The Liberal Tradition in America.*

Chapter 8: Habermas' New Science

1. "Dogmatism, Reason, and Decision," pp. 269–75.

2. "Theorie der Gesellschaft," pp. 152–54.

3. Ibid., p. 145.

4. McCarthy, CTJH, pp. 222–23.

5. "Theorie der Gesellschaft," pp. 266–67.

6. Ibid., p. 276.

7. *Legitimation Crisis*, p. 84.

8. "Theorie der Gesellschaft," p. 237.

9. "Historical Materialism and the Development of Normative Structures," p. 120.

10. Ibid., pp. 97–98.

11. I rely heavily upon Michael Schmid's "Habermas's Theory of Social Evolution."

12. "Historical Materialism and Normative Structures," p. 120.

13. Ibid., p. 121; Schmid, p. 167.

14. "Toward a Reconstruction of Historical Materialism," p. 154; quoted by Schmid, p. 168.

15. "Toward a Reconstruction," p. 160; quoted by Schmid, p. 169.

16. "Toward a Reconstruction," pp. 153–56.

17. It is Martin Jay who notes that Habermas avoids all these traps, in "Habermas and the Reconstruction of Marxist Holism," a paper presented to the Conference Group on the Political Economy of Advanced Industrial Societies (September 1981), pp. 35–37. See too Jay's recent *Marxism and Totality: The Adventures of a Concept from Lukács to Habermas*, pp. 487–95. See also Habermas' "Toward a Reconstruction of Historical Material-

ism," p. 172, and "Historical Materialism and Normative Structures," pp. 120–23.

18. "Historical Materialism," p. 121.
19. Ibid., pp. 98, 121.
20. Schmid, pp. 175–80.
21. Ibid., p. 178.
22. Schmid, p. 180. "Geschichte und Evolution," p. 250. This essay was translated as "History and Evolution" by D. J. Parent. The passage to which Schmid refers to is on p. 29 of the translation.
23. Schmid, p. 180.
24. McCarthy, "Rationality and Relativism," p. 66.
25. Ibid, p. 69.
26. Ibid., p. 72.
27. "What is Universal Pragmatics?", p. 16; quoted by Hesse, "Science and Objectivity," p. 111.
28. "What is Universal Pragmatics?", esp. p. 21.
29. Ibid., p. 16.
30. Ibid., pp. 22, 25. McCarthy, CTJH, pp. 295–98.
31. "What is Universal Pragmatics?", p. 16.
32. McCarthy, CTJH, p. 193.
33. For example, Schmid's criticism in *Critical Debates* seems more stringent than any of that collected in an earlier book of critical debates, *Materialien zu Habermas' 'Erkenntnis und Interesse,'* ed. Winfried Dallmayr.
34. See, for example, Susan Buck-Morss, "Socio-economic Bias in Piaget's Theory and its Implications for Cross-cultural Studies." Cited by McCarthy, "Rationality and Relativism," p. 288, n. 28.
35. Hesse, "Science and Objectivity," p. 112.
36. Habermas, "Toward a Theory of Communicative Competence," pp. 130–32. Noted by Hesse, p. 113.
37. "A Reply," pp. 277–78.
38. *Theorie* 1: 199. *Theory*, p. 138; McCarthy's translation.
39. *Theorie* 1: 199–201. *Theory*, pp. 138–40.
40. *Theorie* 1: 200. *Theory*, p. 139.
41. McCarthy, CTJH, p. 353.
42. "Wozu noch Philosophie?", p. 34.
43. *Theorie* 2: 555.
44. Ibid. 1: 17–18; 161–63; 504–9. *Theory*, pp. 2–4; 108–111; 376–82. Richard Bernstein, in *Beyond Objectivism and Relativism*, suggests that the term "post-empiricist philosophy of science" is an especially apt characterization of the work of these authors. The term, he notes, is Mary Hesse's (pp. 31–33).
45. *Theorie* 1: 504; 161–63. *Theory*, pp. 376–77; 108–11.
46. *Theorie* 1: 161. *Theory*, pp. 376–77.
47. Thomas Kuhn, *The Essential Tension*, p. xiii.

48. Habermas, "Theorie der Gesellschaft oder Sozialtechnologie?", pp. 266–67.

49. *Theorie* 1: 533. *Theory*, pp. 398–99. Translation from a translation of a review of *Theorie* by Johannes Berger, p. 197.

50. Karl Löwith, *Max Weber and Karl Marx*, pp. 57–60. Max Weber, *The Protestant Ethic and the Spirit of Capitalism*, p. 182.

51. *Theorie* 2: 277. Translation from James Schmidt's "Jürgen Habermas and the Difficulties of Enlightenment," p. 191.

52. *Theorie* 1: 299–331. *Theory*, pp. 216–42. *Theorie*, 2: 449–88.

53. "A Reply," pp. 244–45.

54. "Dialectics of Rationalization," p. 28.

55. "Bewusstmachende oder rettende Kritik—Die Aktualität Walter Benjamins," p. 318. Translated as "Consciousness-raising or Pure Critique: The Contemporaneity of Walter Benjamin." Cf. Habermas' "Technology and Science" for his position on Adorno, Horkheimer, and Marcuse's pursuit of a "resurrection of a fallen nature," p. 86.

56. In *Theorie und Praxis*, p. 348. This essay is not included in the English translation, *Theory and Practice*.

57. See Habermas' essay on Adorno, in *Profile*, pp. 182–83.

Chapter 9: Conclusion

1. Ben Agger, in "Marcuse and Habermas on New Science," is an exception. He is impressed with Marcuse's solution.

2. McCarthy, CTJH, pp. 110–36. "Rationality and Relativism," p. 77. Habermas responds in "A Reply," p. 243.

3. "Rationality and Relativism," p. 77. The earlier Habermas source is "Some Difficulties in the Attempt to Link Theory and Praxis," pp. 21–22.

4. Hesse, "Science and Objectivity," p. 106. *Legitimation Crisis*, p. 121.

5. "A Reply," pp. 276–77.

6. Stockman makes a similar point in "Habermas, Marcuse, and the *Aufhebung* of Science and Technology." See p. 35.

7. McCarthy, "Rationality and Relativism," p. 77.

8. Ottmann, "Cognitive Interests and Self-Reflection," p. 89.

9. Ottmann, p. 89. In *The Closing Circle*, Barry Commoner gives a number of examples of how scientific and technological intervention into particular ecological disasters has exacerbated them. See particularly his discussion of eutrophication, chap. 6.

10. Ottmann, p. 89.

11. Ibid.

12. Ibid.

13. Whitebook, p. 66.

14. Whitebook suggests this point.

15. "A Reply," p. 247.
16. Whitebook, p. 52.
17. Ibid., p. 53.
18. Ibid. I presume Whitebook means "worth," and that "work" is a typographical error.
19. Ibid., p. 64.
20. See p. 65, this text.
21. "A Reply," pp. 246–47.
22. Ibid., p. 247.
23. Ibid., p. 241.
24. Ackerman, pp. 127–29.
25. "Some Difficulties in the Attempt to Link Theory and Praxis," pp. 39–40. This is an issue of revolutionary morality (i.e., On what grounds is a small revolutionary group justified in acting in the name of an entire class, or all mankind?), not merely man's relationship to nature.
26. "A Reply," p. 246.
27. Ibid., p. 245.
28. "Psychic Thermidor and the Rebirth of Rebellious Subjectivity," pp. 11–12.
29. "The Classical Doctrine of Politics," pp. 63–64.
30. Maurer, "Habermas' Aufhebung," p. 46.
31. "A Reply," pp. 243–44.
32. Keat, *The Politics of Social Theory*, pp. 78–79.
33. Ibid.
34. Pp. 78–81, this text.
35. P. 92, this text.
36. *Theorie* 1: 106. *Theory*, p. 69. Habermas, "The Entwinement of Myth and Enlightenment: Re-Reading *Dialectic of Enlightenment*," p. 19.
37. *Theorie* 1: 516. *Theory*, p. 385.
38. Habermas, "Die Philosophie als Platzhalter und Interpret," pp. 10, 24–25.
39. "A Reply," p. 249; Habermas' emphasis.
40. Ibid.
41. *Theorie* 2: 583–588.
42. *Ways of Worldmaking*, p. 129.
43. "A Reply," pp. 244, 249; *Theorie* 1: 324, 326. *Theory*, pp. 237–38.
44. "A Reply," pp. 248–49.
45. "A Postscript," pp. 171–72.
46. "A Reply," p. 250; Habermas' emphasis.
47. Maurer, "Habermas' Aufhebung," p. 46.
48. Seyla Benhabib, "Modernity and the Aporias of Critical Theory," pp. 58–59. Murray Bookchin, "Finding the Subject: Notes on Whitebook and 'Habermas Ltd.,'" pp. 97–98.
49. Benhabib, pp. 58–59.
50. Barbara Lex, in "Voodoo Death: New Thoughts on an Old Explana-

tion," reviews the anthropological literature on the subject. David Landy, "Role Adaptation: Traditional Curers Under the Impact of Western Medicine," p. 223. Landy cites several studies of the hand tremblers.

51. Claude Lévi-Strauss, *Structural Anthropology*, vol. 1, chap. 9. William Morgan, "Navaho Treatment of Sickness: Diagnosticians."

52. "A Reply," pp. 243–50.

53. Habermas, "The Classical Doctrine of Politics," pp. 79–81.

54. Habermas, *Knowledge and Human Interests*, chaps. 10–12. Habermas clarifies his position on psychoanalysis as a paradigm of the "emancipatory cognitive interest," in "A Postscript," pp. 182–85.

55. Ottmann, "Cognitive Interests and Self-Reflection," p. 86.

56. Keat, *The Politics of Social Theory*, pp. 96, 107. Christopher Nichols, in "Science or Reflection: Habermas on Freud," makes a similar, but somewhat more measured, argument.

57. David Rasmussen, in a review of *Theorie* in *Philosophy and Social Criticism*, notes that Habermas seems to hold a rather stereotyped view of myth and primitive societies (pp. 8–9). A review of the relevant pages in *Theorie* seems to support this claim. See 1: 72–113. *Theory*, pp. 43–74. McCarthy makes a similar argument in "Rationality and Relativism" p. 69.

58. Landy, "Role Adaptation," pp. 223–24.

59. Lola Romanucci-Ross, "The Hierarchy of Resort in Curative Practices: The Admirality Islands, Melanesia."

60. Benhabib, "Modernity and the Aporias of Critical Theory," pp. 43–44.

61. *Theorie* 1: 512. *Theory*, pp. 382–83.

62. *Theorie* 1: 511–12. *Theory*, p. 382; McCarthy's translation.

63. Adorno, "Die Aktualität der Philosophie," p. 341; quoted and translated by Susan Buck-Morss, *The Origin of Negative Dialectics*, p. 86. I follow closely Buck-Morss' analysis of mimesis on pp. 72–88. Jay, in *Adorno*, discusses the complexity of mimesis, pp. 155–58, also showing that it is constructive, not merely receptive.

64. Adorno, *Kierkegaard: Konstruktion des Aesthetischen*, p. 142; quoted and translated by Buck-Morss, *The Origin of Negative Dialectics*, p. 73.

65. Habermas, "The Entwinement of Myth and Enlightenment: Re-Reading *Dialectic of Enlightenment*," p. 29.

66. Habermas, "Dialectics of Rationalization," p. 8.

67. *Theorie*, 1: 523; my translation. *Theory*, p. 390.

68. Whitebook, "Nature in Habermas," p. 55, fn. 38.

69. *Theorie* 1: 509–10. *Theory*, pp. 380–81; McCarthy's translation.

70. "Psychic Thermidor." p. 9.

71. *Theorie* 1: 514. *Theory*, p. 384.

72. "Psychic Thermidor," p. 9.

73. *Theorie* 2: 561–62, 583.

74. "A Reply," p. 277.

75. Rorty, *Philosophy and the Mirror of Nature*, p. 377.

76. Ibid., p. 382.
77. Ibid., pp. 383–84.
78. See esp. p. 180, in Schmid.
79. Rorty, p. 384.
80. Ibid., pp. 385–86.
81. Paper presented to the Conference on "Hermeneutics and Critical Theory," at Bryn Mawr College, Pennsylvania, February 19, 1983. See esp. pp. 1–7. In a conversation after his presentation, Professor Habermas expanded upon his paper. I draw upon this conversation as well as the paper.
82. "Historical Materialism and Normative Structures," p. 97.
83. "A Reply," p. 221; Bernstein, p. 191.
84. This is also Bernstein's analysis in *Beyond Objectivism and Relativism*, p. 194.
85. Roderick Chisholm, "What is a Transcendental Argument?"
86. Rorty, pp. 388–89.
87. Frank Manuel and Fritzie Manuel, in *Utopian Thought in the Western World*, discuss these and other utopian authors in such a way that Marcuse's continuity with them is apparent. See esp. pp. 1–29. Indeed, Marcuse is included among their utopians (pp. 794–800).
88. *Eros and Civilization*, pp. 146–47.
89. *Bulfinch's Mythology*, by Thomas Bulfinch (N.Y.: Avenel Books, 1979), pp. 101–5, 185–89.
90. Whitebook, pp. 56–59. Stockman, pp. 21ff. Tronn Overend, in "Enquiry and Ideology: Habermas' Trichotomous Conception of Science," criticizes Habermas' conception of science along similar not unconventional lines.
91. Ben Agger is one of the few more attracted to Marcuse's solution than to Habermas'.
92. "A Reply," p. 243.
93. Colletti, *From Rousseau to Lenin*, p. 130.
94. Colin Turnbull, *The Forest People*, esp. chaps. 1, 13, 15.
95. "Technology and Science," p. 99.
96. "The Classical Doctrine," p. 51; my emphasis.
97. Ibid., p. 53.
98. That Habermas does not drift into an authoritarianism of the "real will" is convincingly argued by Stephen White, in "Reason and Authority in Habermas." However, see my reply, in which I argue that White does not frame the issue quite right, in *American Political Science Review*.
99. For the term "*naiv*," see the original "Die klassische Lehre von der Politik," p. 59.

WORKS CITED

Listed here are the German or English editions cited in the text and notes.

Works by Herbert Marcuse

The Aesthetic Dimension: Toward a Critique of Marxist Aesthetics. Boston: Beacon Press, 1978.

"Beiträge zu einer Phänomenologie des Historischen Materialismus." *Philosophische Hefte* 1 (1928): 45–68. In *Herbert Marcuse: Schriften*, vol. 1.

"Contributions to a Phenomenology of Historical Materialism." *Telos*, no. 4 (Fall 1969): 3–34. (Translation of "Beiträge zu einer Phänomenologie des Historischen Materialismus")

Counterrevolution and Revolt. Boston: Beacon Press, 1972.

"Der deutsche Künstlerroman." In *Herbert Marcuse: Schriften* 1:7–346. Frankfurt a.M.: Suhrkamp, 1978.

"The End of Utopia." In *Five Lectures: Psychoanalysis, Politics, and Utopia*, translated by Jeremy Shapiro and Shierry Weber, pp. 62–82. Boston: Beacon Press, 1970.

Eros and Civilization: A Philosophical Inquiry into Freud. New York: Random House, Vintage Books edition, 1962.

An Essay on Liberation. Boston: Beacon Press, 1969.

Hegels Ontologie und die Theorie der Geschichtlichkeit. 3d ed. Frankfurt a.M.: Vittorio Klostermann, 1975.

"Industrialization and Capitalism in the Work of Max Weber." In *Negations: Essays in Critical Theory*, translated by Jeremy Shapiro et al., pp. 201–26. Boston: Beacon Press, 1968.

"Love Mystified: A Critique of Norman O. Brown." In *Negations: Essays in Critical Theory*, translated by Jeremy Shapiro et al., pp. 227–43. Boston: Beacon Press, 1968.
"The Obsolescence of the Freudian Concept of Man." In *Five Lectures: Psychoanalysis, Politics, and Utopia*, translated by Jeremy Shapiro and Shierry Weber, pp. 56–101. Boston: Beacon Press, 1970.
"On Hedonism." In *Negations: Essays in Critical Theory*, translated by Jeremy Shapiro et al., pp. 159–200. Boston: Beacon Press, 1968. (Translation of "Zur Kritik des Hedonismus")
"On Science and Phenomenology." In Boston Studies in the Philosophy of Science, edited by Robert Cohen and Marx Wartofsky, 2:279–90. New York: Humanities Press, 1965.
"On the Philosophical Foundation of the Concept of Labor in Economics." Translated by Douglas Kellner. *Telos*, no. 16 (Summer 1973): 9–37. (Translation of "Über die philosophischen Grundlagen des wirtschaftswissenschaftlichen Arbeitsbegriffs")
"On the Problem of the Dialectic." Translated by Morton Schoolman and Duncan Smith. *Telos*, no. 27 (Spring 1976): 12–39 (includes parts I and II). (Translation of "Zum Problem der Dialektik")
One-Dimensional Man: Studies in the Ideology of Advanced Industrial Society. Boston: Beacon Press, 1964.
"Philosophie und kritische Theorie." In *Kultur und Gesellschaft* 1: 102–27. Frankfurt a.M.: Suhrkamp, 1965.
"Philosophy and Critical Theory." In *Negations: Essays in Critical Theory*, translated by Jeremy Shapiro et al., pp. 134–58. Boston: Beacon Press, 1968. (Translation of "Philosophie und kritische Theorie")
"Das Problem der geschichtlichen Wirklichkeit: Wilhelm Dilthey." *Die Gesellschaft*, vol. 1, no. 4 (1931): 350–67. In *Herbert Marcuse: Schriften*, vol. 1.
"Progress and Freud's Theory of Instincts." In *Five Lectures: Psychoanalysis, Politics, and Utopia*, translated by Jeremy Shapiro and Shierry Weber, pp. 28–43. Boston: Beacon Press, 1970.
"The Realm of Freedom and the Realm of Necessity: A Reconsideration." *Praxis* (Zagreb, Yugoslavia) 5 (1969): 20–25.
"Protosocialism and Late Capitalism." Translated by Michael Vale, Annemarie Feenberg, and Erica Sherover Marcuse. *International Journal of Politics* 10 (Summer-Fall 1980): 25–48.
Reason and Revolution. Boston: Beacon Press, 1960.
"The Responsibility of Science." In *The Responsibility of Power*, edited by L. Krieger and F. Stern, pp. 439–44. Garden City, N.Y.: Doubleday, 1967.
"Sartre's Existentialism." Translated by Joris de Bres. In *Studies in Critical Philosophy*, pp. 157–90. Boston: Beacon Press, 1973.
"Some Social Implications of Modern Technology." *Studies in Philosophy and Social Science* 9 (1941): 414–39.
"The Struggle against Liberalism in the Totalitarian View of the State." In

Negations: Essays in Critical Theory, translated by Jeremy Shapiro et al., pp. 3–42. Boston: Beacon Press, 1968.
"Über die philosophischen Grundlagen des wirtschaftswissenschaftslichen Arbeitsbegriffs." In *Kultur und Gesellschaft*, 2: 7–48. Frankfurt a.M.: Suhrkamp, 1965. In *Herbert Marcuse: Schriften*, vol. 1.
"Zum Problem der Dialektik," I, II (1930, 1931). *Herbert Marcuse: Schriften* 1: 407–55. Frankfurt a.M.: Suhrkamp, 1978.
"Zur Kritik des Hedonismus." In *Kultur und Gesellschaft* 2: 128–68. Frankfurt a.M.: Suhrkamp, 1965.

Works by Jürgen Habermas

"The Analytical Theory of Science and Dialectics." In *The Positivist Dispute in German Sociology*, by Theodor Adorno et al., translated by Glyn Adey and David Frisby, pp. 131–61. New York: Harper & Row, Harper Torchbooks, 1976.
"Bewusstmachende oder rettende Kritik—Die Aktualität Walter Benjamins." In *Kultur und Kritik: Verstreute Aufsätze*, pp. 302–44. Frankfurt a.M.: Suhrkamp, 1973.
"The Classical Doctrine of Politics in Relation to Social Philosophy." In *Theory and Practice*, translated by John Viertel, pp. 41–81. Boston: Beacon Press, 1973. (Translation of "Die klassische Lehre von der Politik in ihrem Verhältnis zur Sozialphilosophie")
"Consciousness-raising or Pure Critique: The Contemporaneity of Walter Benjamin." *New German Critique*, no. 17 (1979): 30–59. (Translation of "Bewusstmachende oder rettende Kritik—Die Aktualität Walter Benjamins")
"Dialectics of Rationalization: An Interview." *Telos*, no. 49 (Fall 1981): 5–31.
"Dogmatism, Reason, and Decision: On Theory and Practice in Our Scientific Civilization." In *Theory and Practice*, translated by John Viertel, pp. 253–82. Boston: Beacon Press, 1973.
"The Entwinement of Myth and Enlightenment: Re-Reading the *Dialectic of Enlightenment*." Translated by Thomas Levin. *New German Critique*, no. 26 (1982): 13–30.
Erkenntnis und Interesse, mit einem neuen Nachwort. Frankfurt a.M.: Suhrkamp, 1977.
"Geschichte und Evolution." *Zur Rekonstruktion des Historischen Materialismus*, pp. 200–59. Frankfurt a.M.: Suhrkamp, 1976.
"Hannah Arendts Begriff der Macht." In *Politik, Kunst, Religion*, pp. 103–26. Stuttgart: Reclam, 1978.
"Hannah Arendt's Communications Concept of Power." Translated by Thomas McCarthy. *Social Research* 44 (1977): 3–24. (Translation of "Hannah Arendts Begriff der Macht")

"Historical Materialism and the Development of Normative Structures." In *Communication and the Evolution of Society*, translated by Thomas McCarthy, pp. 96–129. Boston: Beacon Press, 1979.

"History and Evolution." Translated by D. J. Parent. *Telos*, no. 39 (1979): 5–44. (Translation of "Geschichte und Evolution")

"Die klassiche Lehre von der Politik in ihrem Verhältnis zur Sozialphilosophie." In *Theorie und Praxis: Sozialphilosophische Studien*, pp. 48–88. Frankfurt a.M.: Suhrkamp, 1978.

Knowledge and Human Interests. Translated by Jeremy Shapiro. Boston: Beacon Press, 1971. (Translation of *Erkenntnis und Interesse*)

"Können komplexe Gesellschaften eine vernünftige Identität ausbilden?" In *Zur Rekonstruktion des Historischen Materialismus*, pp. 92–126. Frankfurt a.M.: Suhrkamp, 1976.

Legitimation Crisis. Translated by Thomas McCarthy. Boston: Beacon Press, 1975.

"Moral Development and Ego Identity." In *Communication and the Evolution of Society*, translated by Thomas McCarthy, pp. 70–94. Boston: Beacon Press, 1979.

"Öffentlichkeit (ein Lexikonartikel)." In *Kultur und Kritik: Verstreute Aufsätze*, pp. 61–69. Frankfurt a.M.: Suhrkamp, 1973.

"On Social Identity." *Telos*, no. 19 (1974): 91–103. (Abridged English version of "Können komplexe Gesellschaften eine vernünftige Identität ausbilden?")

Philosophical-Political Profiles. Translated by Frederick G. Lawrence. Cambridge: MIT Press, 1983. (Translation of *Philosophisch-politische Profile*)

"Die Philosophie als Platzhalter und Interpret." In *Moralbewusstsein und kommunikatives Handeln*, pp. 9–28. Frankfurt a.M.: Suhrkamp, 1983.

"A Positivistically Bisected Rationalism." In *The Positivist Dispute in German Sociology*, by Theodor Adorno et al., translated by Glyn Adey and David Frisby, pp. 198–225. New York: Harper & Row, Harper Torchbooks, 1976.

"A Postscript to *Knowledge and Human Interests*." *Philosophy of the Social Sciences* 3 (1973): 157–89.

"Praktische Folgen des wissenschaftlich-technischen Fortschritts." In *Theorie und Praxis: Sozialphilosophische Studien*, pp. 336–58. Frankfurt a.M.: Suhrkamp, 1978.

"Psychic Thermidor and the Rebirth of Rebellious Subjectivity." *Berkeley Journal of Sociology* 24–25 (1980): 1–12.

"The Public Sphere: An Encyclopaedia Article." Translated by Sara Lennox and Frank Lennox. *New German Critique*, no. 3 (1974): 49–55. (Translation of "Öffentlichkeit")

"A Reply to my Critics." In *Habermas: Critical Debates*, edited by John B. Thompson and David Held, pp. 219–83. Cambridge: MIT Press, 1982.

"The Scientization of Politics and Public Opinion." In *Toward a Rational Society*, translated by Jeremy Shapiro, pp. 62–80. Boston: Beacon Press,

1970. (Translation of "Verwissenschaftlichte Politik und öffentliche Meinung")

"Some Difficulties in the Attempt to Link Theory and Praxis." In *Theory and Practice*, translated by John Viertel, pp. 1–40. Boston: Beacon Press, 1973.

Strukturwandel der Öffentlichkeit. Neuwied and Berlin: Luchterhand, 1962.

"Technical Progress and the Social Life-World." In *Toward a Rational Society*, translated by Jeremy Shapiro, pp. 50–61. Boston: Beacon Press, 1970.

"Technik und Wissenschaft als 'Ideologie.'" In *Technik und Wissenschaft als 'Ideologie'*, pp. 48–103. Frankfurt a.M.: Suhrkamp, 1968.

"Technology and Science as 'Ideology.'" In *Toward a Rational Society*, translated by Jeremy Shapiro, pp. 81–122. Boston: Beacon Press, 1970. (Translation of "Technik und Wissenschaft als 'Ideologie'")

"Theodor W. Adorno: Ein philosophierender Intellektueller." In *Philosophisch-politische Profile*, pp. 176–84. Frankfurt a.M.: Suhrkamp, 1971.

"Theodor W. Adorno: Urgeschichte der Subjektivität und verwilderte Selbstbehauptung." In *Philosophisch-politische Profile*, pp. 184–99. Frankfurt a.M.: Suhrkamp, 1971.

With Niklas Luhmann. "Theorie der Gesellschaft oder Sozialtechnologie? Eine Auseinandersetzung mit Niklas Luhmann." In *Theorie der Gesellschaft oder Sozialtechnologie—Was leistet die Systemforschung?*, pp. 142–290. Frankfurt a.M.: Suhrkamp, 1971.

Theorie des kommunikativen Handelns, 2 vols. Frankfurt a.M.: Suhrkamp, 1981.

The Theory of Communicative Action. Translated by Thomas McCarthy. Boston: Beacon Press, 1984. (Translation of *Theorie des kommunikativen Handelns*, vol. 1)

"Toward a Reconstruction of Historical Materialism." In *Communication and the Evolution of Society*, translated by Thomas McCarthy, pp. 130–77. Boston: Beacon Press, 1979.

"Toward a Theory of Communicative Competence." In *Recent Sociology*, edited by Hans Peter Dreitzel, no. 2, pp. 115–48. New York: Macmillan Co., 1970.

"Der Universalitätsanspruch der Hermeneutik." In *Kultur und Kritik: Verstreute Aufsätze*, pp. 264–301. Frankfurt a.M.: Suhrkamp, 1973.

"Verwissenschaftlichte Politik und öffentliche Meinung." In *Technik und Wissenschaft als 'Ideologie.'* Frankfurt a.M.: Suhrkamp, 1968.

With Niklas Luhmann. "Vorbereitende Bemerkungen zu einer Theorie der kommunikativen Kompetenz." In *Theorie der Gesellschaft oder Sozialtechnologie—was leistet die Systemforschung?*, pp. 101–41. Frankfurt a.M.: Suhrkamp, 1971.

"Wahrheitstheorien." In *Wirklichkeit und Reflexion: Walter Schulz zum sechzigsten Geburtstag*, edited by H. Fahrenbach. Pfullingen: Neske, 1973.

"What is Universal Pragmatics?" In *Communication and the Evolution of So-*

ciety, translated by Thomas McCarthy, pp. 1–68. Boston: Beacon Press, 1979.

"Why More Philosophy?" *Social Research* 38 (1971): 633–54. (Translation of "Wozu noch Philosophie?")

"Wozu noch Philosophie?" In *Philosophisch-politische Profile*, pp. 11–36. Frankfurt a.M.: Suhrkamp, 1971.

"Zu Nietzsches Erkenntnistheorie (ein Nachwort)." In *Kultur und Kritik: Verstreute Aufsätze*, pp. 239–63. Frankfurt a.M.: Suhrkamp, 1973.

Zur Logik der Sozialwissenschaften. Frankfurt a.M.: Suhrkamp, 1970.

Other Works

Ackerman, Bruce. *Social Justice in the Liberal State*. New Haven: Yale University Press, 1980.

Adorno, Theodor. "Die Aktualität der Philosophie." In *Frühe philosophische Schriften*, edited by Rolf Tiedemann, pp. 325–44. Frankfurt a.M.: Suhrkamp, 1973. This is vol. 1 of his *Gesammelte Schriften*.

———. *Kierkegaard: Konstruktion des Aesthetischen*. Tübingen: Verlag J. C. B. Mohr, 1966.

———. "Subject and Object," pp. 497–511. In *The Essential Frankfurt School Reader*, edited by Andrew Arato and Eike Gebhardt. New York: Urizen Books, 1978. (This essay originally appeared in Adorno's *Stichworte*, 1969.)

Agger, Ben. "Marcuse and Habermas on New Science." *Polity* 9 (1976): 158–81.

Albert, Hans. "Hermeneutik und Realwissenschaft." In *Kritische Vernunft und menschliche Praxis*, pp. 127–79. Stuttgart: Reclam, 1977.

———. "The Myth of Total Reason: Dialectical Claims in the Light of Undialectical Criticism." In *The Positivist Dispute in German Sociology*, by Theodor Adorno et al., translated by Glyn Adey and David Frisby, pp. 163–97. New York: Harper & Row, 1976.

Alford, C. Fred. "Comment on 'Reason and Authority in Habermas.'" *American Political Science Review* 75 (June 1981): 463–64.

———. "Jürgen Habermas and the Dialectic of Enlightenment: What is Theoretically Fruitful Knowledge?" *Social Research* (forthcoming).

———. "Review of Kātz's *Herbert Marcuse and the Art of Liberation*." *Journal of Politics* 46 (August 1984): 981–83.

Apel, Karl-Otto. *Towards a Transformation of Philosophy*. Translated by Glyn Adey and David Frisby. London: Routledge & Kegan Paul, 1980.

———. "Wissenschaft als Emanzipation?" In *Materialien zu Habermas' 'Erkenntnis und Interesse,'* edited by Winfried Dallmayr, pp. 318–48. Frankfurt a.M.: Suhrkamp, 1974.

Arendt, Hannah. *The Human Condition*. Chicago: University of Chicago Press, 1958.

Ashton, T. H. *The Industrial Revolution, 1760–1830*. London: Oxford University Press, 1948.

Bacon, Francis. *The Great Instauration and New Atlantis*. Edited by J. Weinberger. Arlington Heights, Ill.: Harlan Davidson, 1980.

Bahro, Rudolf. *The Alternative in Eastern Europe*. Translated by David Fernbach. London: New Left Books, 1978.

Benhabib, Seyla. "Modernity and the Aporias of Critical Theory." *Telos*, no. 49 (1981): 39–59.

Berger, Johannes. "Review of *Theorie des kommunikativen Handelns*." Translated by David J. Parent. *Telos*, no. 57 (1983): 194–205.

Berndt, Heide and Reimut Reiche. "Die geschichtliche Dimension des Realitätsprinzips." In *Antworten auf Herbert Marcuse*, edited by Jürgen Habermas, pp. 104–33. Frankfurt a.M.: Suhrkamp, 1968.

Bernstein, Richard. *Beyond Objectivism and Relativism: Science, Hermeneutics, and Praxis*. Philadelphia: University of Pennsylvania Press, 1983.

———. *Praxis and Action*. Philadelphia: University of Pennsylvania Press, 1971.

Böhler, Dietrich. "Zur Geltung des emanzipatorischen Interesses," in *Materialien zu Habermas' 'Erkenntnis und Interesse,'* edited by Winfried Dallmayr, pp. 349–68. Frankfurt a.M.: Suhrkamp, 1974.

Bookchin, Murray. "Finding the Subject: Notes on Whitebook and 'Habermas Ltd.'" *Telos*, no. 52 (1982): 78–98.

Brown, Harold I. "Incommensurability." *Inquiry* 26 (1983): 3–29.

Buck-Morss, Susan. *The Origin of Negative Dialectics*. New York: The Free Press, 1977.

———. "Socio-economic Bias in Piaget's Theory and its Implications for Cross-cultural Studies." *Human Development* 18 (1975): 35–49.

Chisholm, Roderick. "What is a Transcendental Argument?" *Neue Heft für Philosophie* 14 (1978): 19–22.

Colletti, Lucio. *From Rousseau to Lenin*. Translated by John Merrington and Judith White. London: New Left Books, 1972.

Commoner, Barry. *The Closing Circle*. New York: Bantam Books, 1974.

Conant, James B. *Science and Common Sense*. New Haven: Yale University Press, 1951.

D'Amico, Robert. "Review of William Leiss' *The Domination of Nature*." *Telos*, no. 15 (Spring 1973): 142–47.

Feyerabend, Paul. "Against Method." Minnesota Studies in the Philosophy of Science, edited by Herbert Feigl et al., 4: 17–130. Minneapolis: University of Minnesota Press, 1970.

———. *Against Method*. London: New Left Books, 1975.

———. *Philosophical Papers*, 2 vols. London: Cambridge University Press, 1981.

———. *Science in a Free Society*. London: New Left Books, 1978.

Freud, Sigmund. *Civilization and its Discontents*. Translated by James Strachey. New York: W. W. Norton, 1961.

Fromm, Erich. *Escape From Freedom*. New York: Avon Books, 1965.

Galston, William. "Review of *Social Justice in the Liberal State*." *Political Theory* 9 (1981): 427–33.

Giddens, Anthony. "The Improbable Guru: Re-Reading Marcuse." In *Profiles and Critiques in Social Theory*, pp. 144–63. Berkeley and Los Angeles: University of California Press, 1982.

———. "Labor and Interaction." In *Habermas: Critical Debates*, edited by John B. Thompson and David Held, pp. 149–61. Cambridge, Mass.: MIT Press, 1982.

Glaser, William. *Soziales und instrumentales Handeln*. Stuttgart: Kohlhammer, 1972.

Goodman, Nelson. *Ways of Worldmaking*. Indianapolis: Hackett Publishing Co., 1979.

Gurwitsch, Aron. "Comment on the Paper by H. Marcuse." Boston Studies in the Philosophy of Science, vol. 2, edited by Robert Cohen and Marx Wartofsky, pp. 291–306. New York: Humanities Press, 1965.

Hartz, Louis. *The Liberal Tradition in America*. New York: Harcourt Brace Jovanovich, 1955.

Haug, Wolfgang Fritz. "Das Ganz und das ganz Andere. Zur Kritik der reinen revolutionären Transzendenz." In *Antworten auf Herbert Marcuse*, edited by Jürgen Habermas, pp. 50–72. Frankfurt a.M.: Suhrkamp, 1968.

Held, David. *Introduction to Critical Theory*. Berkeley and Los Angeles: University of California Press, 1980.

Hesse, Mary. "Habermas's Consensus Theory of Truth." In *Proceedings of the 1978 Biennial Meeting of the Philosophy of Science Association*, vol. 2, edited by P. D. Asquith and I. Hacking. East Lansing, Michigan, 1978.

———. "Science and Objectivity." In *Habermas: Critical Debates*, edited by John B. Thompson and David Held, pp. 98–115. Cambridge, Mass.: MIT Press, 1982.

Hessen, Boris. "The Social and Economic Roots of Newton's *Principia*." In *The Rise of Modern Science*, edited by G. Basalla, pp. 31–38. Lexington, Mass.: Heath, 1968.

Horkheimer, Max. *Eclipse of Reason*. New York: Seabury Press, 1974.

———. "Traditional and Critical Theory." In *Critical Theory*, translated by Matthew J. O'Connell, et al., pp. 188–252. New York: Seabury Press, 1972.

Horkheimer, Max and Theodor Adorno. *Dialectic of Enlightenment*. Translated by John Cumming. New York: Herder and Herder, 1972.

Husserl, Edmund. *The Crisis of European Sciences and Transcendental Phenomenology*. Translated by David Carr. Evanston, Ill.: Northwestern University Press, 1970.

Jay, Martin. *Adorno*. Cambridge, Mass.: Harvard University Press, 1984.

———. *The Dialectical Imagination*. Boston: Little, Brown & Co., 1973.

———. *Marxism and Totality: The Adventures of a Concept from Lukács to*

Habermas. Berkeley and Los Angeles: University of California Press, 1984.

Kātz, Barry. *Herbert Marcuse and the Art of Liberation: An Intellectual Biography*. London: New Left Books, Verso editions, 1982.

Keat, Russell. *The Politics of Social Theory: Habermas, Freud, and the Critique of Positivism*. Chicago: University of Chicago Press, 1981.

Kellner, Douglas. *Herbert Marcuse and the Crisis of Marxism*. Berkeley and Los Angeles: University of California Press, 1984.

Kohlberg, Lawrence. "From Is to Ought." In *Cognitive Development and Epistemology*, edited by T. Mishel. New York: Academic Press, 1971.

Koyré, Alexandre. *From the Closed World to Infinite Universe*. Baltimore: Johns Hopkins Press, 1957.

Krüger, Lorenz. "Überlegungen zum Verhältnis wissenschaftlicher Erkenntnis und gesellschaftlicher Interessen." In *Materialien zu Habermas' 'Erkenntnis und Interesse,'* edited by Winfried Dallmayr, pp. 200–219. Frankfurt a.M.: Suhrkamp, 1974.

Kuhn, Thomas. *The Essential Tension*. Chicago: University of Chicago Press, 1977.

———. *The Structure of Scientific Revolutions*, 2d ed., enlarged. Chicago: University of Chicago Press, 1970.

———. "Theory Change as Structure Change." In *Historical and Philosophical Dimensions of Logic, Methodology, and Philosophy of Science*, edited by R. Butts and J. Hintikka. Dordrecht: D. Reidel, 1977.

Lakatos, Imre. "Falsification and the Methodology of Scientific Research Programmes." In *Criticism and the Growth of Knowledge*, edited by Imre Lakatos and Alan Musgrave, pp. 91–196. Cambridge University Press, 1970.

Landy, David. "Role Adaptation: Traditional Curers Under the Impact of Western Medicine." In *Health and the Human Condition*, edited by Michael Logan and Edward Hunt, Jr., pp. 217–41. North Scituate, Mass.: Duxbury Press, 1978.

Lasch, Christopher. *The Minimal Self: Psychic Survival in Troubled Times*. New York: W. W. Norton & Co., 1984.

Leiss, William. *The Domination of Nature*. Boston: Beacon Press, 1972.

———. "The Problem of Man and Nature in the Work of the Frankfurt School." *Philosophy of the Social Sciences* 5 (1975): 163–72.

———. "Technological Rationality: Marcuse and His Critics." *Philosophy of the Social Sciences* 2 (1972): 31–42.

Lévi-Strauss, Claude. *Structural Anthropology*, vol. 1. New York: Basic Books, 1967.

Lex, Barbara. "Voodoo Death: New Thoughts on an Old Explanation." *American Anthropologist* 76 (1974): 818–23.

Lobkowicz, Nikolaus. "Interest and Objectivity." *Philosophy of the Social Sciences* 2 (1972): 193–209.

Lowi, Theodor. *The End of Liberalism*. New York: W. W. Norton & Co., 1969.

Löwith, Karl. *Max Weber and Karl Marx*. Edited by Tom Bottomore and William Outhwaite. Translated by Hans Fantel. London: George Allen and Unwin, 1982.

Macaulay, Thomas Babington. "Lord Bacon." *Critical and Historical Essays*, edited by F. C. Montague, 2: 115–240. London: Methuen and Co., 1903.

MacIntyre, Alasdair. *Herbert Marcuse*. New York: Viking Press, 1970.

Malinovich, Myriam Miedzian. "On Herbert Marcuse and the Concept of Psychological Freedom." *Social Research* 49 (1982): 158–80.

Manuel, Frank and Fritzie Manuel. *Utopian Thought in the Western World*. Cambridge: Belknap Press of Harvard, 1979.

Maurer, Reinhart. "Jürgen Habermas' Aufhebung der Philosophie." *Philosophische Rundschau*, Beiheft 8 (1977).

McCarthy, Thomas. *The Critical Theory of Jürgen Habermas*. Cambridge: MIT Press, 1978.

———. "Rationality and Relativism: Habermas's 'Overcoming' of Hermeneutics." In *Habermas: Critical Debates*, edited by John B. Thompson and David Held, pp. 57–78. Cambridge, Mass.: MIT Press, 1982.

———. "A Theory of Communicative Competence." *Philosophy of the Social Sciences* 3 (1973): 135–56.

McConnell, Grant. *Private Power and American Democracy*. New York: Random House, 1966.

Merton, Robert K. *Science, Technology, and Society in Seventeenth-Century England*. New York: H. Fertig, 1970.

More, Thomas. *Utopia*. Edited by Edward Surtz. New Haven: Yale University Press, 1964.

Morgan, William. "Navaho Treatment of Sickness: Diagnosticians." *American Anthropologist* 33 (1931): 390–402.

Mousnier, Roland. *Progrès scientifique et technique au XVIIIe siècle*. Paris: Pion, 1958.

Nagel, Ernest. *The Structure of Science*. New York: Harcourt, Brace, and World, 1961.

Needham, Rodney. Introduction to *Primitive Classification*, by Emile Durkheim and Marcel Mauss. Chicago: University of Chicago Press, 1963.

Nichols, Christopher. "Science or Reflection: Habermas on Freud." *Philosophy of the Social Sciences* 2 (1972): 261–70.

Offe, Claus. "Technik und Eindimensionalität: Eine Version der Technokratiethese?" In *Antworten auf Herbert Marcuse*, edited by Jürgen Habermas, pp. 73–88. Frankfurt a.M.: Suhrkamp, 1968.

Ottmann, Henning. "Cognitive Interests and Self-Reflection." In *Habermas: Critical Debates*, edited by John B. Thompson and David Held, pp. 79–97. Cambridge: MIT Press, 1982.

Overend, Tronn. "Enquiry and Ideology: Habermas' Trichotomous Conception of Science." *Philosophy of the Social Sciences* 8 (1978): 1–13.

Piaget, Jean. *Genetic Epistemology*. Translated by Eleanor Duckworth. New York: W. W. Norton & Co., 1971.

Popper, Karl. *The Open Society and Its Enemies*, 5th ed. revised, vol. 1. Princeton: Princeton University Press, 1966.

———. "Three Views Concerning Human Knowledge." In *Conjectures and Refutations*, pp. 97–119. New York: Harper & Row, 1965.

Primack, Joel and Frank von Hippel. *Advice and Dissent*. New York: New American Library, 1974.

Rasmussen, David. "Review of *Theorie des kommunikativen Handelns*." *Philosophy and Social Criticism* 9 (1982): 1–28.

Romanucci-Ross, Barbara. "The Hierarchy of Resort in Curative Practices: The Admirality Islands, Melanesia." *Journal of Health and Social Behavior* 10 (1969): 201–09.

Rorty, Richard. "In Defense of Eliminative Materialism." *The Review of Metaphysics* 24 (September 1970): 112–21.

———. *Philosophy and the Mirror of Nature*. Princeton: Princeton University Press, 1979.

———. *Science, Technology, and Human Values* 5 (Winter 1980): pp. 43–44.

Schmid, Michael. "Habermas's Theory of Social Evolution." In *Habermas: Critical Debates*, edited by John B. Thompson and David Held, pp. 162–80. Cambridge: MIT Press, 1982.

Schmidt, Alfred. *The Concept of Nature in Marx*. Translated by Ben Fowkes. London: New Left Books, 1971.

———. "Existential-Ontologie und historischer Materialismus bei Herbert Marcuse." In *Antworten auf Herbert Marcuse*, edited by Jürgen Habermas, pp. 17–49. Frankfurt a.M.: Suhrkamp, 1968.

Schmidt, James. "Jürgen Habermas and the Difficulties of Enlightenment." *Social Research* 49 (1982): 181–208.

Schoolman, Morton. *The Imaginary Witness: The Critical Theory of Herbert Marcuse*. New York: The Free Press, 1980.

Schroyer, Trent. *The Critique of Domination*. Boston: Beacon Press, 1975.

Senghaas, Dieter. "Sozialkybernetik und Herrschaft." In *Texte zur Technokratiediskussion*, edited by Claus Koch and Dieter Senghaas, pp. 196–216. Frankfurt a.M.: Europäische Verlagsanstalt, 1970.

Stockman, Norman. "Habermas, Marcuse and the *Aufhebung* of Science and Technology." *Philosophy of the Social Sciences* 8 (1978): 15–35.

Strauss, Leo. "Plato." In *Political Philosophy*, edited by Hilail Gildin, pp. 159–237. New York: Bobbs-Merrill Co., Pegasus Books, 1975.

———. "What is Political Philosophy?". In *Political Philosophy*, edited by Hilail Gildin, pp. 3–57. New York: Bobbs-Merrill Co., Pegasus Books, 1975.

Toulmin, Stephen. *Human Understanding*, vol. 1. Princeton: Princeton University Press, 1972.

Turnbull, Colin. *The Forest People: A Study of the Pygmies of the Congo*. New

York: Simon and Schuster, Touchstone Books, 1961.

Wartofsky, Marx. *Conceptual Foundations of Scientific Thought*. London: Macmillan Co., 1968.

Weber, Max. *The Protestant Ethic and the Spirit of Capitalism*. London: George Allen and Unwin, 1976.

Wellmer, Albrecht. *Critical Theory of Society*. Translated by John Cumming. New York: Seabury Press, 1974.

White, Alan R. "Coherence Theory of Truth." *The Encyclopedia of Philosophy*, vol. 2, pp. 130–33. New York: Macmillan Co., 1967.

White, Stephen. "Reason and Authority in Habermas." *American Political Science Review* 74 (December 1980): 1006–19.

Whitebook, Joel. "The Problem of Nature in Habermas." *Telos*, no. 40 (Summer 1979): 41–69.

INDEX